T0230696

Systems-Level Modelling of Microbial Communities

Theory and Practice

Focus Computational Biology Series

This series aims to capture new developments in computational biology and bioinformatics in concise form. It seeks to encourage the rapid and wide dissemination of material for emerging topics and areas that are evolving quickly. The titles included in the series are meant to appeal to students, researchers, and professionals involved in the field. The inclusion of concrete examples and applications, and programming techniques and examples, is highly encouraged.

Systems-Level Modelling of Microbial Communities

Theory and Practice

Aarthi Ravikrishnan and Karthik Raman

Systems-Level Modelling of Microbial Communities

Theory and Practice

Aarthi Ravikrishnan
Karthik Raman

CRC Press
Taylor & Francis Group
Boca Raton London New York

CRC Press is an imprint of the
Taylor & Francis Group, an **informa** business

CRC Press
Taylor & Francis Group
6000 Broken Sound Parkway NW, Suite 300
Boca Raton, FL 33487-2742

© 2019 by Taylor & Francis Group, LLC
CRC Press is an imprint of Taylor & Francis Group, an Informa business

No claim to original U.S. Government works

Printed and bound by CPI Group (UK) Ltd, Croydon, CR0 4YY
Version Date: 20180808

International Standard Book Number-13: 978-1-138-59671-9 (Hardback)

Library of Congress Cataloging-in-Publication Data

Names: Ravikrishnan, Aarthi, author.
Title: Systems-level modelling of microbial communities : theory and practice / Aarthi Ravikrishnan, Karthik Raman.
Description: Boca Raton, Florida : CRC Press, [2019] | Includes bibliographical references and index.
Identifiers: LCCN 2018030137| ISBN 9781138596719 (hardback : alk. paper) | ISBN 9780429487484 (ebook)
Subjects: LCSH: Microbial ecology--Mathematical models. | Microbial ecology--Methodology.
Classification: LCC QR100 .R38 2019 | DDC 579/.17--dc23
LC record available at https://lccn.loc.gov/2018030137

Visit the Taylor & Francis Web site at
http://www.taylorandfrancis.com

and the CRC Press Web site at
http://www.crcpress.com

To my mom, dad and sister

—Aarthi Ravikrishnan

To my parents and teachers

—Karthik Raman

Contents

Preface

From the Great Barrier Reef in Australia to a biofilm that forms on an implant in the body, microbial communities are ubiquitous. In the human gut, they outnumber the human cells and contribute as a key metabolic organ in human growth and development. Microbial communities have far-reaching implications in human health and disease; they also have an immense potential for biotechnological applications. This book provides a glimpse of how *in silico* models can throw light on how microbes interact and thrive in communities.

The recent years have witnessed an increasing interest in the development of both experimental and computational methods to study and characterise microbial communities. This book outlines many of these methods. Specifically, we focus on modelling interactions between the microbes in a community with special emphasis on network-based and constraint-based modelling techniques. We also briefly discuss experimental methods to understand/characterise these microbial communities.

This book is fundamentally aimed at researchers in the field of computational and systems biology as well as biologists and experimentalists studying microbial communities, who are keen on embracing the concepts of computational modelling. There are many exciting challenges in modelling microbial communities *in silico* — this book but scratches the surface of this burgeoning field. Nevertheless, we hope that it can pique the interest of more scientists and attract them to the exciting field of microbial community modelling.

Chennai   *Aarthi Ravikrishnan*
                                                       Karthik Raman

CHAPTER 1

Introduction to microbial communities

Micro-organisms are ubiquitous, yet rarely function as independent entities, and thrive in *communities*. Landscapes of microbial communities are enormous and display astounding capabilities to perform composite actions. The dynamic and complex interactions between microbes play significant roles in shaping community assemblies. Microbial communities are inarguably the pre-eminent friends and foes of human beings, influencing human health and disease, as well as the biosphere itself.

Interactions in microbial communities can be broadly classified into beneficial, neutral or harmful, which determine the spatio-temporal arrangement of the micro-organisms [1]. Such complex interactions help microbes inhabit different niches, which otherwise would not have been possible. For instance, in the biofilm formed on the surface of the teeth, the initial colonisers bind to the salivary pellicle receptors, followed by the later colonisers, which attach to the receptors of initial micro-organisms [2].

Microbial communities have also evoked a lot of interest for metabolic engineering. Owing to the diversity in the metabolic capabilities of the constituent micro-organisms, microbial communities are emerging as a viable alternative to single organisms. Naturally occurring microbial consortia have been explored for several applications such as wastewater treatment, food fermentations and bioremediation. Further, many of the beneficial associations between micro-organisms have been leveraged in multiple industrial applications, where co-cultures of organisms with diverse metabolic capabilities have been used. In addition, metabolic interactions between microbes have been synthetically induced by introducing heterologous pathways and transporters. These transporters serve as a gateway for interactions between the micro-organisms [3].

Due to the advantages discussed above and the amenability of microbial communities to *manipulations*, there is a burgeoning interest in applying these communities for biotechnological applications, despite the associated practical difficulties. Further, to better understand microbial communities, several modelling techniques have been developed, each of which provides a different perspective. In this chapter, we shine a spotlight on the basic concepts underlying microbial communities and discuss examples elucidating the importance of microbial communities.

1.1 MODES OF INTERACTION IN COMMUNITIES

Micro-organisms in a community tend to interact with each other through various means. These interactions can be broadly classified into different categories, based on the nature of interactions. One such type of interaction is mutualism, where all the organisms benefit each other. A classic example of mutualism includes syntrophy, where there is a cross-feeding of the nutrients between the micro-organisms in a community. Such cross-feeding of nutrients has been commonly observed in nature, especially in anaerobic marine environments and under extreme conditions of temperature and pH [4].

Syntrophic associations have also been demonstrated in the laboratory, where two organisms — *Desulfovibrio vulgaris* Hildenborough and *Methanococcus maripaludis* S2 — were co-cultivated in a medium containing lactate [5]. Further, this co-culture system was also modelled to understand the mutualistic interactions [6]. In another example, it was shown that strains of *Methanobacillus omelianskii*, which are the most abundant organisms in sewage sludge, show cross-feeding tendencies when grown on ethanol [7].

In another type of interaction, commensalism, one of the organisms derives benefit, while there is no effect on the other organisms in the community. Such an interaction is most commonly found in co-cultivation of lactic acid bacteria (LAB) and propionic acid bacteria (PAB), where the lactic acid produced by the former serves as the carbon source for the latter. The LAB, however, does not derive any benefit from the presence of PAB [8]. In the human gut, although the organisms are often referred to as commensals, the relationship between the gut microbiome and the host is predominantly mutualistic [9].

Other types can be broadly categorised under *harmful* interactions, where either one or all of the organisms are negatively affected. Parasitism is one such form of interaction, where there is a benefit for one organism (parasite) at the cost of the other (host). Such an interaction is most commonly observed between bacteria–bacteriophage or between any predator and prey [10]. In the other type of interaction, amensalism, one of the organisms is harmed, while the other derives no benefit from the interaction. Such interactions are most commonly observed in food fermentations, especially those involving lactic acid bacteria [11], where the end products reduce the pH of the fermentation medium, detrimentally affecting the other organisms in the community. Competition is one other type of interaction, where the organisms compete with each other for space and resources.

In general, while microbial communities are used for any application, positive or *beneficial* interactions are generally leveraged, so that the joint metabolic capabilities of the microorganisms can be fully explored.

Key Interactions

- **Mutualism (+/+)**: All organisms in the community are benefited

- **Commensalism (+/0)**: Only one organism is benefited, others derive no benefit

- **Amensalism (-/0)**: One organism derives no benefit while the others are harmed

- **Competition (-/-)**: All the organisms compete for same resources, and the community comes to harm

- **Parasitism (-/+)**: Only one organism is benefited, others are harmed

In the next sections, we mention a few examples of well-known naturally existing microbial communities and also describe a few studies where synthetic microbial communities have been used for biotechnological applications.

1.2   NATURAL MICROBIAL COMMUNITIES

Several microbial communities are known to inhabit different natural habitats, such as the soil, the human oral cavity and the gut. In many communities, micro-organisms act in tandem, carrying out a specific function, while in a few other cases, such as communities in biofilms, there is a sequential progression or succession of micro-organisms. The naturally existing microbial communities have implications for multiple processes, ranging from the digestion of human milk oligosaccharides [12] to the regulation of global biogeochemical cycles [13]. Microbial communities also play important roles in the progression of several human diseases, including diabetes mellitus and atherosclerosis [14]. In this section, we describe a few important naturally existing microbial communities, and the roles played by the micro-organisms therein.

1.2.1   Gut microbiota

The human gut is known to be inhabited by a distinctive community of micro-organisms. The composition of the human gut microbiota is known to vary at different stages of the lifecycle. These variations are influenced by several factors such as perinatal colonisation, diet, host susceptibility, exposure to antibiotics and other environmental exposures [15]. Several studies [16, 17] have shown that the composition of the gut microbiota is primarily influenced by the mode of birth and the type of diet. These studies have reported that the gut of infants born via vaginal delivery is dominated by *Lactobacillus* and *Prevotella* species, while those born via Caesarean delivery have abundant *Clostridium*, *Staphylococcus*, *Propionobacterium* and *Corynebacterium* species [16]. Moreover, the development of these micro-organisms in the gut of infants depends on the mode of feeding — in breastfed infants, it was shown that a few species of *Bifidobacterium* such as *B. infantis*, *B. breve*, *B. adolescentis*, *B. longum* and *B. bifidum* are abundant, in comparison to formula-fed infants, who had higher levels of *B. longum* [17].

Since the organisms in the gut are involved in a plethora of functions, there has been an emerging interest in analysing and understanding the gut microbiota. Specifically, many studies focus on the interplay of organisms and other factors such as lifestyle and dietary intake [18]. The gut microbiota often aid in the conversion of indigestible dietary polysaccharides to short-chain fatty acids such as acetate, propionate and butyrate, which are then absorbed by the human host cells [19]. Other prominent functions of the gut micro-organisms include the development of immune system, metabolism of drug and xenobiotics, as well as protection from resident pathogens [20].

Recent advances in gene sequencing technologies combined with modelling techniques have enabled the understanding of these microbial communities and also their interactions. Specifically, metabolic modelling of microbial communities has gained a lot of traction [21,22] partly due to an increase in the availability of genomic data. Further, to gather more profound insights into the metabolic capabilities of the gut, genome-scale metabolic reconstructions have been developed for several gut microbiota [23]. More details on the genome-scale modelling of gut microbial communities can be found in refs. [24,25].

1.2.2   Soil microbiota

Soil is inhabited by different types of microbial communities, which help in carbon sequestration and maintaining the fertility of the soil. The micro-organisms in the soil also regulate the biogeochemical cycles [26] and play an important role in several processes such as decomposition of organic matter found in forest soils [27].

The soil has different horizons, or layers, each supporting unique sets of micro-organisms. The composition of the soil microbiota varies with the type and the pH of soil [28]; they have profound effects on the growth of the plants, conferring them with the ability to withstand abiotic stress, and providing resistance against several plant pathogens.

Several studies have shown a bi-directional association between the composition of the soil microbiome and the micro-organisms found on the plants [29,30]. In one such study, soil microbiome was depicted to be one of the most influential factors in determining the microbiome associated with the aerial parts of the grapevine plant [29]. This study reported that the microbial communities found on the leaves and the flowers belong to the taxa most commonly found in the soil. In another study [30], the starch yields in the tuberous roots of sweet potatoes were correlated to the microbial communities found in the soil rhizosphere. Several genera of bacteria such as *Sphingobium*, *Pseudomonas* and *Acinetobacter* were found to be significantly abundant in the rhizosphere of sweet potato genotype with the lowest starch content. However, there was an increased abundance of bacteria belonging to the *Bacillus* genus in the rhizosphere associated with all types of sweet potato tubers.

1.2.3 Microbial communities in biofilms

Biofilms are aggregations of bacterial cells most commonly embedded in a self-synthesised polymer-type of a matrix, which may or may not be attached to biotic or abiotic surfaces. These biofilms are often found in the human body, where they are associated with different types of diseases such as dental caries, melioidosis and cystic fibrosis pneumonia [31]. Due to the collective behaviour and the spatial organisation of different populations of micro-organisms, biofilms display several unique properties.

The co-operation between the co-aggregated organisms in the biofilms, such as that observed between *Streptococci* and *Actinomyces*, often aids in inhabiting different surfaces such as the enamel surfaces of teeth [32]. Besides, several reasons underlie the formation of biofilms, each offering their unique set of ecological advantages. The complex matrix in which the bacteria co-aggregate, comprises a mixture of components such as proteins, extracellular polysaccharides and nucleic acids, secreted by the constituent micro-organisms. This matrix, commonly referred to as the extracellular polymeric substance (EPS) matrix, plays a crucial role in protecting the micro-organisms, by acting as a barrier to the entry of specific anti-microbial agents [33]. Moreover, the EPS matrix also protects the bacteria from a wide variety of environmental stresses such as changes in pH or osmotic conditions, and desiccation [34]. For instance, it was observed that the biofilm composed of *Pseudomonas aeruginosa*, *Pseudomonas protegens* and *Klebsiella pneumoniae*, exhibited an increased resistance towards the anti-microbials sodium dodecyl sulphate (SDS) and tobramycin, in comparison to the respective mono-cultures [35]. Similarly, in another study [36], it was found that the biofilms formed by *Zymomonas mobilis* display an enhanced tolerance towards benzaldehyde, in comparison to free-living planktonic cells.

The spatial organisation of the micro-organisms in the biofilm gives rise to several *emergent properties*. The proximity of micro-organisms not only promotes the mutual exchange of metabolites (syntrophy), but also aids in horizontal gene transfer leading to the acquisition of new traits. Further, the structure of biofilms allows the penetration of nutrients and promotes the stable co-existence of different types of bacteria. Due to these advantages, microbial biofilms have industrial applications ranging from bioremediation [37] to the production of fine chemicals [36,38].

Traditionally, microbial biofilms have been most commonly used in wastewater treatment processes, especially in trickle-bed bioreactors. Here, the process of biofilm formation by a population of diverse micro-organisms is initiated on inert surfaces; the biofilm eventually degrades and reduces the organic content in the wastewater [39]. The application of biofilms has also been demonstrated for the production of industrially useful chemicals such as poly(3-hydroxybutyrate) [38].

1.3   SYNTHETIC MICROBIAL COMMUNITIES

Microbial communities, due to their enhanced capabilities, are better suited for several biotechnological applications. A consortium enjoys the privilege of *division of labour* and provides a wider scope to leverage the joint metabolic capabilities of the organisms.

Mixed populations of microbes have been used in the past, particularly in wastewater treatments and food fermentations, where the end goal is the overall conversion of the organic content [40]. However, while designing a well-defined consortium for any application, different paradigms are possible, as illustrated in Figure 1.1. Most of these strategies seek to improve either mutualism or commensalism to enhance the stability of the consortium. On this basis, four different strategies, namely sequential utilisation, co-utilisation, substrate transformation and product transformation, were proposed, each of which can give rise to stable consortia [41].

Over the recent years, there has been an emerging interest in using these microbial communities for industrial applications. Synthetic microbial communities, tailored for a specific purpose, have been used in the production of biofuels, and speciality chem-

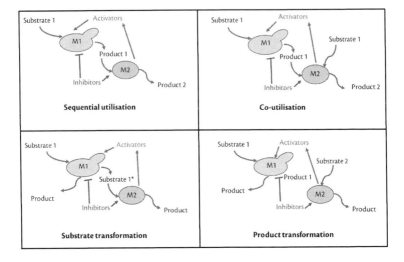

Figure 1.1 **Strategies for designing a microbial consortium.** Four different strategies for designing synthetic microbial communities as described
in [41]. Sequential utilisation, where there is a series of metabolite transfer, one organism synthesising a product, which is used as the substrate
by the other. Substrate transformation and product transformation can be
broadly classified under this category. In case of the former, one organism
renders the substrate in a useful form, which can be used by the second
organism. In the latter, the second organism uses the by-product of the
first organism and converts it to produce the desired product of interest. Co-utilisation is another category, where the organisms use the same
substrate for producing different products. However, this type of interaction introduces competition between the organisms. The *activators* and
inhibitors include a pool of compounds consisting of various molecules,
including metabolites and quorum sensing moieties.

icals including vitamin C. In all these applications, a population of micro-organisms with a known set of metabolic characteristics have been used. Further, to exploit the fullest potential of microbial communities and to reduce the operational costs, several process optimisations have been carried out. One classic example is the development of *Consolidated Bioprocessing (CBP)* paradigm for the production of ethanol from plant biomass, which is rich in lignin, cellulose and hemi-cellulose. Early methods to convert lignocellulose to ethanol involved separate stages of enzymatic hydrolysis and fermentation, each carried out by dedicated sets of micro-organisms. However, these processes were not economically favourable, due to lower yields and higher capital [42]. CBP, in contrast, is a one-step process, which employs a microbial consortium with collective capabilities for substrate utilisation and product formation [43]. Due to this single-stage fermentation, the process cost was significantly reduced [42].

In another study to examine the production of cellulose from ethanol, a one-pot bioreactor was designed using two different strains, *Acremonium cellulolyticus* C-1 and *Saccharomyces cerevisiae*. This study reported a significant improvement in the overall rate of ethanol production, indicating that such a single-stage fermentation is scalable for large-scale conversion of cellulose to ethanol [44]. Many other studies also report improved ethanol production using mixed populations of microbes [45, 46].

Well-defined synthetic communities have also been used for the production of other chemicals. Traditionally, vitamin C is produced from glucose using a two-step fermentation process, where a community of micro-organisms is used for the conversion of L-sorbose to 2-keto-L-gulonic acid (2-KLG, a precursor of vitamin C). This well-defined community consists of *Ketogulonicigenium vulgare* and *Bacillus megaterium*, where the former produces 2-KLG, while the latter is a helper strain enhancing the growth of the former. The metabolomic and proteomic analysis carried out on this community revealed both mutualistic and antagonistic behaviours. Specifically, it was observed that *B. megaterium* secretes purine substrates and nutrients, which enhances the growth of *K. vulgare* [47]. However, *K. vulgare*, despite secreting amino acids, inhibits the growth of *B. megaterium* [48].

The use of synthetic microbial communities with genetically engineered micro-organisms has also gained a lot of traction [49, 50]. Here, individual micro-organisms engineered with specialised pathways and transporters are grown together for the production of desired metabolites. Such a strategy not only makes use of the metabolic capabilities of multiple organisms, but also alleviates the increased burden of introducing foreign pathways in individual organisms. For instance, in one such study [49], two strains of *Escherichia coli*, BuT-8L-ato and BuT-3E, were engineered with two distinct metabolic pathways to synthesise *n*-butanol. The former harbours genes to synthesise butyrate from acetate, while the latter has genes to synthesise *n*-butanol from butyrate. These two organisms interact with one another through the exchange of acetate and butyrate.

In another study [50], 3-aminobenzoic acid (3AB) was produced using a co-culture of *E. coli* cells, engineered with distinct biosynthetic pathways. The entire pathway was divided into two modules – upstream module, producing the precursor 3-dehydroshikimic acid (DHS), and the downstream module converting DHS to 3AB. This co-culture system was shown to have 15-fold higher 3AB production, compared to the mono-culture reconstituted with a complete 3AB biosynthetic pathway. Further, *E. coli–E. coli* co-culture systems have also been used for the production of a commodity chemical, *cis,cis*-muconic acid, from feedstocks containing a mixture of glucose and xylose. This co-culture system was shown to have several advantages over traditional mono-cultures including high yield and better efficiency [3].

1.4   MODELLING MICROBIAL COMMUNITIES

Naturally-existing and synthetic microbial communities have been widely modelled to understand the general mechanisms that drive micro-organisms to stay together as a community. Several well-designed frameworks to design and understand microbial dynamics have been developed using genomic, proteomic and metabolomic data. Often, microbial communities are modelled to understand the species abundances, population dynamics, spatial distributions and interactions between them in terms of the cross-feeding potential and the competition for resources. The modelling techniques employed are of different types, each tailored to analyse a specific type of data.

We classify these methods broadly into network-based, population-based, individual-based and constraint-based techniques depending on the level of abstraction, nature of problem formulation and the type of application. A schematic illustrating the steps of network-based and constraint-based approaches can be found in Figure 1.2.

Network-based methods employ a graph-based representation of microbial communities. The *microbial association* graphs or networks are constructed from metagenomic data. These networks are then analysed through a series of pipelines and techniques to infer the types of relationships between the microorganisms. Besides, network-based methods have also been used to study the metabolic interactions between the organisms. We discuss in detail about these network-based modelling techniques in Chapter 2.

Several studies, as previously discussed in §1.2.1, §1.2.2 and §1.2.3 have pointed towards the collective role of individual organisms in shaping a community. The structural organisation of microbial communities is often dictated by the population dynamics and the underlying interactions. Such paradigms are well captured in *population-based* and *individual-based models*. These modelling techniques, as discussed in Chapter 3, are useful in understanding population dynamics in light of spatial and temporal relationships.

The other class, collectively known as *constraint-based modelling techniques* are useful in understanding the metabolism of microbial communities. The knowledge of metabolic interactions is essential while using microbial communities for industrial applications. The techniques under this category seek to understand the metabolic phenotype and determine the type of interactions between the organisms. For instance, Stolyar *et al* [6] demonstrated the mutualistic interactions in a community consisting of *Desulfovibrio vulgaris* and *Methanococcus maripaludis* under steady state conditions. In addition, there have also been efforts to include dynamic data into these steady state constraint-based approaches. In this case, in addition to the usual paradigm of solving linear equations, Ordinary Differential Equations (ODEs) that describe the rate of substrate uptake and product formation are first solved. These values are then assigned to corresponding reactions, and flux balance analysis is carried out on the resultant model. Further details about these techniques can be found in Chapter 4.

Microbial communities have been extensively modelled using one or more of the above techniques to discern the underlying complex interactions. A broad overview of a few studies under every category can be found in Table 1.1. Each of these studies

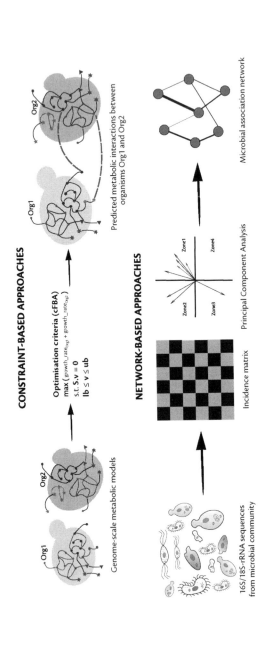

Figure 1.2 **Illustration of different techniques to model microbial communities:** Top panel shows the general steps involved in the constraint-based modelling of microbial communities. The input to this technique is the genome-scale metabolic models of the micro-organisms. These models are then analysed through different linear programming based approaches to predict the *carrying capacity* of microbial communities and the type of interactions between the micro-organisms. The bottom panel describes the network-based approaches, where the most commonly used input data are genomic sequences representing different phyla. The information from these sequences are then processed, and different types of statistical procedures are applied to determine the microbial association networks. These networks are further analysed through different techniques.

TABLE 1.1 Some recent advances in modelling microbial communities.

Method type	Description	Ref.
CbM	Modelling syntrophic interactions between *G. metallireducens* and *G. sulfurreducens* using genome-scale metabolic models	[51]
CbM	Development of a combined metabolic model of two organisms, *K. vulgare* and *B. megaterium*, to demonstrate the production of vitamin C	[52]
CbM	Analysis of interactions of bacteria in human gut by developing genome-scale models of three key gut microbes	[24]
CbM	Development of a mixed-integer linear programming (MILP) based method to quantify the interactions between micro-organisms in a minimal medium	[22]
CbM, NbM	Graph-theoretic approach to identify interactions in a co-culture of genetically modified *E. coli*	[53]
NbM	Microbiome modelling using metabolic network and relative abundance data to understand the correlation between species co-occurence and metabolic interactions	[21]
NbM	Analysis of co-occurrence networks of soil micro-organisms in different microbiomes	[54]
NbM	Host-pathogen interaction measured using a graph-based algorithm that uses the data from metabolic network	[55]
NbM	Development of a dynamic-programming based algorithm to enumerate biosynthetic pathways in metabolic communities	[56]
IbM	Development of BacSim to model the behaviour of micro-organisms using individual properties	[57]
PbM	Understanding co-operations during range expansions using partial differential equations	[58]

CbM: Constraint-based methods, NbM: Network-based methods, IbM: Individual-based methods, PbM: Population-based methods provides a different lens to study microbial communities and obtain interesting insights into their architecture and function.

Key points

- Micro-organisms are ubiquitous and exist as communities where they exhibit different forms of interactions

- Natural microbial communities are found to occupy different niches, ranging from human gut to soil

- Due to their enriched metabolic capabilities, synthetic microbial consortia have been used for many applications

- Different types of modelling techniques have been developed to understand the emergent properties in microbial communities

---

FURTHER READING

- Zengler K and Palsson BØ (2012) A road map for the development of community systems (CoSy) biology. *Nat Rev Microbiol* **10**(5):366–372

- Biggs MB, Medlock GL, Kolling GL, and Papin JA (2015) Metabolic network modeling of microbial communities. *Wiley Interdiscip Rev Syst Biol Med* **7**(5):317–334

- Bosi E, Bacci G, Mengoni A, and Fondi M (2017) Perspectives and challenges in microbial communities metabolic modeling. *Front Genet* **8**:88

- D'Souza G, Shitut S, Preussger D, Yousif G, Waschina S, and Kost C (2018) Ecology and evolution of metabolic cross-feeding interactions in bacteria. *Nat Prod Rep* **35**(5):455–488

CHAPTER 2

Network-based modelling of microbial communities

Principles of network science or graph theory have been extensively applied to several domains, including sociology, computer science and ecology to address several interesting questions. Network-based theories and principles have also been extended to study biological processes such as predicting the role of proteins from their interaction networks, identifying potential genes that serve as drug targets and understanding signalling cascades.

Over the recent years, understanding microbial communities using concepts from network science has gained much traction. This partly owes to the tremendous increase in the amount of genomic or metagenomic sequences arising from a myriad of microbial communities. Networks of communities are often constructed using such metagenomic data, in addition to the information available from the metabolic networks of the microbes.

Often, networks are modelled as graphs, where the nodes represent different entities, and the edges depict the interactions or relationships between these nodes. These networks or graphs are analysed using different types of graph-theoretic algorithms to understand the underlying relationships. Moreover, multiple types of statistical analyses can be performed using these networks to quantify several properties such as the relative importance of each node and the strength of interactions. In this chapter, we discuss the fundamentals of networks and elaborate on a few network-based modelling techniques, which have been used to quantify and understand the relationships between the microorganisms.

2.1 GRAPH THEORY FUNDAMENTALS

Graphs are defined by a set of nodes or vertices (V) connected to each other through edges ($E \subseteq V \times V$) representing the relationships between the nodes. Different types of graphs exist, which can be defined as follows:

1. **Undirected graph**: A graph $G(V, E)$ is undirected, if the edge between two nodes A and B, $(A, B) \in E$ is the same as $(B, A) \in E$.

2. **Directed graph**: A graph $G(V, E)$ is directed, if the edge between two nodes A and B, $(A, B) \in E$ is not the same as $(B, A) \in E$.

3. **Weighted graph**: A graph $G(V, E)$ is weighted, if the edge between two nodes A and B, $(A, B) \in E$ has particular attributes or weights associated with it.

4. **Bi-partite graph**: An undirected graph $G(V, E)$ is bi-partite, if V can be partitioned into two sets V_1 and V_2 such that $(u, v) \in E$ implies either $u \in V_1$ and $v \in V_2$ or $u \in V_2$ and $v \in V_1$ [59].

5. **Hypergraph**: An undirected graph $G(V, E)$ is a hypergraph, where every *hyperedge* connects a set of vertices $V' \subseteq 2^V$ and not just two of them.

6. **Cyclic graph**: A directed graph $G(V, E)$ is a cyclic graph, if there is a path $\langle v_0, v_1, \ldots, v_k \rangle v_i \in V$ such that $v_0 = v_k$.

2.2 MICROBIAL ASSOCIATION NETWORKS

Recent advances in metagenomic sequencing have led to a deeper understanding of microbial communities. Experimental techniques, especially those pertaining to sequencing, generate a large

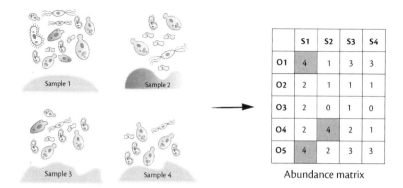

	S1	S2	S3	S4
O1	4	1	3	3
O2	2	1	1	1
O3	2	0	1	0
O4	2	4	2	1
O5	4	2	3	3

Abundance matrix

Figure 2.1   Schematic to illustrate a data matrix of microbial communities. Microbes from different samples are isolated, sequenced and assigned to their respective taxa depending on the conserved regions. The sequence reads are then used to quantify the microbes belonging to specific taxa and represented in the form of an abundance matrix as shown.

quantum of data, which can be systematically analysed to derive valuable insights (also see Chapter 5). From the metagenomic data, it is now possible to quantify the relative abundances and infer the relationships between the organisms in a community. Multiple pipelines have been developed, which make use of the sequence reads and render this information as species abundance. As a general routine, reads from the sequences are analysed, and extensive text mining is carried out to assign these sequences to the respective taxa. These data are often represented in the form of a *binary matrix* indicating the presence or absence of the species in the given sample. Besides, the relative abundances of the organisms can also be determined by counting the number of reads and enlisting them in the form of a data matrix, where every row represents the read count of the species in every sample (column), as illustrated in Figure 2.1.

The association networks are then constructed using a collection of techniques referred to as *network inference,* where the relationships between the microbes are inferred based on the abundance data. These techniques can be broadly classified into two major types, one predicting the pairwise relationships and the other analysing the cumulative effect of multiple species over one of them [10].

2.2.1   Predicting pairwise interactions

Pairwise interactions, which depict the co-occurrence or co-exclusion of two species, are one of the simplest forms of interactions to predict and understand. Such pairwise relationships convey a lot of information about the mutual dependence or competition between the organisms and help quantify the distribution of the two species. One such method to quantify these interactions is based on *ecological distance*, calculated based on the similarity or dissimilarity of the species present in a given sample, as quantified by different types of metrics. These indices, namely the Jaccard index, the Bray–Curtis index and the Dice index have been used to quantify the relationships between the organisms [60–63]. The definitions of these can be found below:

1. **Jaccard index**: This index measures the similarity between two samples A and B of finite sizes, and is defined as follows:

$$J(A, B) = \frac{|A \cap B|}{|A \cup B|}$$

   Jaccard index ranges between 0 and 1: zero indicates that the two *sets* A and B are completely different, while one implies that A and B are identical. Jaccard distance, which quantifies the distance between two sets, is defined as $1 - J(A, B)$.

2. **Sørensen index or Dice coefficient**: This coefficient measures the abundance of elements in a pair of samples A and B and is expressed as:

$$Dice(A, B) = \frac{2 \times |A \cap B|}{|A| + |B|}$$

3. **Bray–Curtis dissimilarity index**: This index can be used to quantify the differences in the number of species present in two different samples and is expressed as:

$$BC(A, B) = 1 - \frac{2 \times C_{AB}}{|S_A| + |S_B|}$$

   where A and B are the two samples,
   C_{AB} is sum of the counts of least number of species found in both the samples
   S_A is the total number of organisms in sample A
   S_B is the total number of organisms in sample B.

2.2.2   Understanding higher-order relationships

In addition to the pairwise relationships, it is also important to identify higher-order interactions, where the presence of one species is often affected by the combined abundances of the others. In such cases, the association between microbes is inferred by correlating the abundances of a group of organisms. To this end, several coefficients such as the Spearman correlation coefficient and the Pearson correlation coefficient are used, which quantify the relationships between the microbes. However, many a time, the data from these metagenomic studies are sparsely distributed and tend to be biased towards the sampling areas [64]. Such data could lead to false predictions of the interactions, and thus require significant corrections before making inferences. Many tools have been developed, to remove the bias and address such discrepancies.

One such method, based on regression, called SparCC [65], seeks to overcome such compositional biases by calculating the Pearson correlations between the log-transformed data. This method determines the empirical correlations by making approximations based on the following assumptions: (a) the number of factors under consideration is large, and (b) the average correlation factor between the organisms is small. Another algorithm, called CCREPE (Compositionality Corrected by Permutation and Renormalization) [62], also accounts for the compositional effects by eliminating erroneous correlations. This algorithm operates using two major steps: permutation and bootstrapping. In the first step, a null distribution based on the normalised uncorrelated species data, which indicates the compositional bias, is generated. In the second step, another distribution is generated by bootstrapping over the sample sets and re-calculating the correlation coefficients. This distribution is checked against the null distribution to determine if it is significantly different.

Beyond regression-based methods, there are other classes of techniques that rely on different methodologies, namely Bayesian networks, Local Similarity Analysis (LSA) and graphical model inference. For instance, SPIEC-EASI [66] (SParse InversE Covariance Estimation for Ecological ASsociation Inference), is a statistical method that relies on the graphical representation of the microbial community. This method also assumes a sparse ecological network and seeks to integrate data transformation methods with the graph-based models. Specifically, two procedures, namely the neighbourhood selection and the inverse covariance selection methods are used to infer the underlying ecological network. Further, to account for the time series data, multiple tools employ Local Similarity Analysis (LSA) [67,68] to decipher the dynamics of a biological system. Based on real-time data, this method determines the changes in relative abundance of one species with respect to the other, under different environmental conditions. In addition to these techniques, methods based on Bayesian frameworks have also been developed [69], which integrate the data from relative abundances of microbes and predict the microbial community compositions.

2.3 COMMUNITY METABOLIC NETWORKS

In addition to predicting the relative abundances of microbes, metagenomic sequences also lend themselves to building genome-scale metabolic networks, which can be used to understand the metabolism of microbial communities. The topology of metabolic networks provides several insights, particularly relating to the metabolic capacity and the niche occupied by these organisms. *Community metabolic networks* are constructed by merging the metabolic networks of individual organisms. The microorganisms in the community are assumed to interact with one another through a common extracellular medium. Such networks can further be studied using different types of network-based approaches.

One such technique called *network expansion* [70] can be applied on such metabolic networks to study *biosynthetic capacity*. This process starts with a set of *seed compounds*, which represent the chemical entities found in the environment where the organisms thrive. Starting with these compounds, the network is traversed based on the constituent reactions, expanding the network on the go. At the end of this traversal, a set of metabolites, denoted as *scope* is obtained, which consists of all the metabolites that can be produced using the seed compounds. Based on these principles, co-operation between organisms has been studied [71] by analysing the biosynthetic gain acquired by one organism in the presence of other. Besides these, the topology of metabolic networks has also been used to identify the metabolites that have to be externally supplied for the organism to survive. Such a paradigm, called *reverse ecology* has been widely applied to trace the evolutionary history of the metabolic networks and the environmental niches of microbes [60]. A more detailed description can be found in the following sections.

2.3.1 Identification of seed compounds: the reverse ecology framework

The definition of *seed compounds* in the context of reverse ecology differs slightly from the one mentioned above. Here, *seed compounds* refer to the minimal set of compounds, which (a) cannot be synthesised by the given metabolic network and have to be supplied exogenously, and (b) whose presence helps in the synthesis of all the other compounds in the network. These seed compounds are identified from the metabolic networks by decomposing them into *Strongly Connected Components* (SCC). Each of these components is analysed, and the nodes with no incoming edges and at least one outgoing edge are defined as the *source components*. All these source components, taken together, contribute to the candidate seed compounds. Since every connected component can be viewed as an *equivalence class*, the set of seed compounds includes exactly one node from each of these components. It obviously follows that the candidate set of seed compounds is thus not unique.

Based on this concept of seed metabolites, the interspecies interactions between a pair of organisms from human microbiome was predicted [21]. Specifically, two different indices to quantify the interactions between a pair of organisms (say A and B) were put forth:

Metabolic competition index Fraction of seed metabolites of A that overlap with the seed metabolites of B.

Metabolic complementarity index Fraction of seed metabolites of A that are found in the metabolic network of B but do not overlap with B's seed metabolites.

A strong positive or a negative correlation between the co-occurrence and these indices indicate that the inter-species interactions play a strong role in microbial community assembly. For instance, a strong positive correlation between metabolic competition index and co-occurrence indicates that the organisms tend to compete with each other. However, a similar correlation between metabolic complementarity index and co-occurrence points toward co-operation between the micro-organisms. In addition to these, other indices such as effective metabolic overlap [72] and biosynthetic support score [73] have also been developed, which characterise the level of competition between two species and the nutritional support offered by the host to the parasite, respectively.

Multiple tools have been developed, which enable the determination of seed metabolite set and quantification of metabolic interactions based on the indices mentioned above. Few of these include:

1. *NetSeed* [74], a web-based tool and a Perl module to determine the seed metabolite set.

2. *NetCmpt* [75], a web-based tool and a standalone software package to calculate the effective metabolic overlap [72] and quantify the competition between the organisms.

3. *NetCooperate* [73], a web-based tool and a standalone software package to predict host–microbe and microbe–microbe interactions.

2.3.2 MetQuest: Understanding metabolic exchanges in microbial communities

Community metabolic networks can also be used to determine the biosynthetic pathways producing the metabolites of interest. Graph-based approaches have been developed, which enable the identification of pathways involved in the conversion of a source metabolite to a product of interest. Such a traversal, when extended to a community of organisms (i.e., a community metabolic network) provides insights into the metabolic interactions happening therein. MetQuest [56] is one such method, which identifies all possible biosynthetic pathways between a given set of source and target molecules. This algorithm is based on dynamic programming and efficiently handles large metabolic networks such as those of microbial communities. Here, the metabolic networks are modelled as a directed bi-partite graph $G(M, R, E)$, where M is the set of metabolites in the metabolic network, R is the set of reactions and E is the set of edges.

MetQuest employs two stages — a guided breadth-first search, followed by the assembly of pathways. In the first phase, the joint metabolic network is traversed in a *breadth-first* manner starting from the given input of seed metabolites S. As the first step, the reactions whose reactants are present in seed metabolite set S are identified and marked *visited*. The metabolites produced by these reactions are then added to S, and the reactions where these metabolites participate are added to the queue. Such an expansion continues till there are no reactions that can be *visited*. This phase provides the set of *visited* reactions and the *scope* of seed metabolite set, which consists of all the metabolites that can be produced from the seed metabolite set.

The second phase employs a dynamic programming-based assembly of pathways that produce the target metabolite from the source. The algorithm maintains a *Table*, whose rows and columns correspond to the metabolites and pathway size respectively. Every cell in this table contains set(s) of pathways of a particular size that produces the target metabolites from the source. A particular cell in this *Table* is populated using the entries of pathways from the previous iteration, which are chosen based on appropriate combinations of pathway sizes generated at every step. In the end, the table is filled with all possible pathways till a particular size cut-off is reached. The pathways producing the target metabolite from this table can be further analysed to determine the metabolites getting exchanged between the organisms.

2.4   CASE STUDIES

Network-based approaches have been used in several cases to examine and understand different types of relationships such as those between host–parasites and host–symbionts. As detailed in the previous sections, such modelling can be done either using metagenomic sequence data or metabolic networks. Multiple studies have been carried out along these lines, particularly focussing on the complex interactions between the microbiome found across different sites of the human body. In this aspect, the Human Microbiome Project (HMP) [76,77] has been an invaluable source of metagenomic sequences, providing unprecedented information on the resident micro-organisms. These sequences have been studied and modelled as networks to reveal the types of interactions and relationships between micro-organisms present in different parts of the human body. Further, such modelling can also be used to distinguish the disease-associated changes from the normal healthy states.

In one such study [62], the microbial co-occurrence network was constructed using 726 taxa from HMP to understand the ecological interactions between the organisms. It was observed that this network was scale-free, containing a few hub nodes indicating the relative importance of such organisms. Further, the co-occurrence patterns of the microbes were more often observed to be highly localised to a particular site. Moreover, the organisms that were closer on the phylogenetic tree exhibited a positive association, while those being distant with similar functionalities were found to compete with one another.

In another study [78], phylogenetic and functional annotations were carried out on the metagenomic sequences from human faecal samples of a European population. Different classes of analyses combined with the co-occurrence networks helped in the identification of three major *enterotypes*. These enterotypes are clusters of organisms predominantly belonging to three major genera — *Bacteroides, Prevotella* and *Ruminococcus* — with varying abundances. Similarly, network-based techniques have also been used to differentiate the gut microbiota [79] found in healthy and alcoholic individuals. Specifically, the 16S-rRNA sequences were obtained from the gut of different individuals and the reads were assigned to specific taxa. The co-occurrence of the organisms was modelled as an undirected graph, with nodes as the organisms, and edges running between the nodes depending on the average abundance of organisms in the samples. From this network, different parameters were calculated, and motifs were identified to determine if there were any specific interacting patterns. Interestingly, the differences in gut microbiota prevalent in alcoholic and healthy subjects could be clearly distinguished based on network parameters such as degree distribution, average clustering coefficient and average network diameter.

In addition to the metagenomic sequences, network-based analyses of microbial communities have also been carried out using metabolic networks. For instance, the endosymbiotic relationship between two different species of bacteria, *Sulcia muelleri* and *Baumannia cicadellinicola*, and the insect host *Homalodisca coagulata*, has been studied using the metabolic networks reconstructed from their respective genome sequences [80]. Graph-based analyses on this joint metabolic network indicated a one-way dependence of *B. cicadellinicola* on the metabolism of *S. muelleri* to produce its proteins.

Besides such independent data analyses, integrating time series metagenomic information with genome-scale metabolic networks has also been explored [81]. This type of modelling was used to demonstrate the effects of the antibiotic Clindamycin on the infection caused by *Clostridium difficile* and identify if there were any gut bacteria slowing down its growth. For this purpose, a dynamic network model was constructed from the metagenomic sequences obtained from mouse gastrointestinal tract. Using this model, *Barnesiella intestinihominis* was identified to slow the growth of *C. difficile*. Moreover, this model was also able to capture the dynamics of the antibiotic clindamycin.

Key points

- Network-based analyses on microbial communities rely on the information from metagenomic sequences and metabolic networks

- Metagenomic data has been used to construct microbial association networks to infer the relationships between the organisms

- Metabolic networks of microbial communities have been used to decipher the underlying ecological principles governing the niche preferences

- Metagenomic data can also be integrated with metabolic networks to construct dynamic network models

---

FURTHER READING

- Li C, Lim KMK, Chng KR, and Nagarajan N (2016) Predicting microbial interactions through computational approaches. *Methods* **102**:12–19

- Barabasi AL and Oltvai ZN (2004) Network biology: understanding the cell's functional organization. *Nat Rev Genet* **5**(2):101–113

- Newman MEJ (2011) *Networks: an introduction*. Oxford Univ. Press. ISBN 9780199206650

CHAPTER 3

Population- and agent-based modelling of microbial communities

Microbial community ecosystems comprise ensembles of micro-organisms connected through a complex web of interactions. Ecological succession of microbial communities is one of the most commonly observed natural phenomena. Predominantly, such successions are seen when there are dynamic changes in the environment. These sudden *drifts*, when accumulated over large time-scales, lead to tremendous changes in the composition of the ecosystem.

While network-based methods such as those discussed in Chapter 2 are useful to identify large-scale associations from metagenomics data and even study metabolic exchanges in a community, they often cannot shed light on the dynamic nature of the interactions between microbes in a community. The evolutionary dynamics and the time-course evolution of microbial communities can be better understood using modelling techniques based on ODEs and Partial Differential Equations (PDEs), respectively.

Alternatively, the microbial ecosystems can also be studied using agent-based (individual-based) models, which take into account the individual properties of the micro-organisms and the interactions within the population. In this chapter, we elaborate on population-based and agent-based modelling techniques, throwing light on a few interesting studies carried out on these lines. Figure 3.1 presents an overview of the methods discussed in this chapter.

3.1 POPULATION-BASED MODELLING

Traditionally, microbial ecosystems are modelled at the population level to discern several ecological principles. These models provide an overall idea about the interacting populations. The models under this category predominantly use ODEs or PDEs to describe the underlying characteristics such as the population dynamics and the spatial distribution of micro-organisms at a community level. This class of models also includes *strategy-based techniques*, which are more commonly known as *game-theoretic models*. In this section, we discuss the basic framework underlying each of these methods and elaborate with a few examples.

3.1.1 ODE-based modelling

Models based on ODEs are most widely used to capture the population dynamics of a given ecosystem. ODEs have also been used to understand the interactions between the organisms in a microbial community. The most classic ODE model for modelling population dynamics is the simple yet versatile Lotka–Volterra model [82], which is typically used to capture the dynamics of multiple species in an environment, e.g. a predator and its prey.

3.1.1.1 Lotka–Volterra model

Below, we first describe the basic Lotka–Volterra model, and then go on to generalisations of the same, for capturing microbial community dynamics.

$$\frac{dX}{dt} = \alpha X - \beta XY$$
$$\frac{dY}{dt} = \gamma XY - \delta Y$$

$$(3.1)$$

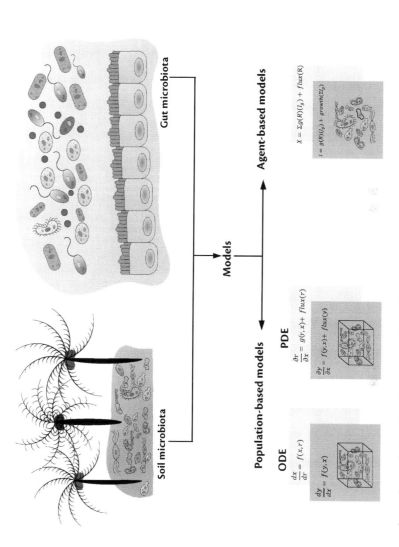

Figure 3.1 **Overview of population-based and agent-based techniques.** Population-based models involve ODEs and PDEs to understand the population and spatial dynamics of microbial communities. Agent-based modelling technique accounts for the properties of individual organisms in the microbial communities.

where X is the number of prey

Y is the number of predators

α is the growth rate (birth rate) of prey

β is the predator's rate of consumption of prey

γ is the growth rate of predator consuming the prey

δ is the death rate of the predator

$\dfrac{dX}{dt}$ and $\dfrac{dY}{dt}$ are the growth rates of prey and predator respectively.

This model is often helpful in predicting interspecific competition and niche differences. The main assumptions of this model include:

1. The prey population increases exponentially with an intrinsic rate, independent of the density of the predators.

2. The predation rate is constant.

3. There is no interference among predators.

4. There is a constant conversion rate of the eaten prey into new predator abundance.

5. There is a constant death rate of predators.

6. The prey has an unlimited food supply.

7. The predator is dependent only on a single species of prey.

8. There is no threat to prey other than the predator.

9. There is no migration of species.

The Lotka–Volterra equations are also generalised to model community dynamics and predict the behaviour of the community under different types of conditions [83, 84]. The generalised Lotka–Volterra equation is given as:

$$\frac{dx_i(t)}{dt} = x_i(t)\left(b_i + \sum_{j=1}^{N} a_{ij}x_j(t)\right) \tag{3.2}$$

$$i = 1, ..., N$$

where

N is the number of different taxa of organisms in the community

$x_i(t)$ is the abundance of taxon i at time t

b_i is the growth rate of taxon i

a_{ij} is the strength of interaction between taxon i and j.

Although such a simple equation does not account for the total abundances of the species or complex processes, it can be extended to build more accurate community models.

3.1.1.2 ODE-based modelling of Quorum Sensing

Quorum Sensing (QS) is a mechanism by which the cells communicate with one another through small molecules known as *autoinducers*. The transcription machinery pertaining to certain genes are activated when they sense the signalling molecules such as N-acyl homoserine lactones, dialkylresorcinols and cyclohexanediones secreted by the other organisms, which in turn regulate several processes such as aggregation, virulence, luminescence and biofilm formation.

Several models have been developed to decipher the processes underlying QS. These models are often based on ODEs and/or PDEs, which are used to understand cell–cell communication from the perspective of dynamic changes in the population and the biochemistry of the QS mechanism. In this section, we restrict to discussing ODE-based modelling of the QS mechanisms.

Early studies [85,86] modelling QS focussed on understanding the biochemistry underlying the signalling cascade. These studies modelled the internal concentrations of QS molecules (QSM), QS proteins (QSP) and the complexes formed by the interaction of QSMs and QSPs as a function of time. This model, when applied to understand the bioluminescence in *Vibro fischeri* [85], predicted two stable steady states, one exhibiting bioluminescence and the other corresponding to non-bioluminescence. In the other study [86] extending this model, the population density of *Pseudomonas aeruginosa* was considered as a parameter, in addition to the internal concentrations of QSMs. This model also predicted a bi-stable steady state that depended on the population density.

Additional details such as the influence of signalling molecules on the regulation of bacterial cell populations were also included in the model to understand the QS mechanism in *V. fischeri* [87]. This model assumes a well-mixed population of both these types of cells. The model, which seeks to capture the cell growth of these populations and the rate of QSM production, can be described as follows:

$$\frac{dN_d}{dt} = r(N_d + (2-\gamma)N_u)F(N_d + N_u) - \alpha G(A)N_d + \beta N_u$$

$$\frac{dN_u}{dt} = r(\gamma - 1)N_u F(N_d + N_u) + \alpha G(A)N_d - \beta N_u$$

$$\frac{dA}{dt} = k_u N_u + k_d N_d - \alpha G(A)N_d - \lambda A$$

(3.3)

where N_d is the number of down-regulated cells per unit volume
N_u is the number of up-regulated cells per unit volume
$r, \alpha, \beta, \gamma, \lambda$ are parameters governing different processes
$F(\cdot)$ is a dimensionless bacterial growth function
$G(\cdot)$ is a function describing the process of QSM-QSP formation
A is the concentration of QSMs.

Further, QS has been extensively studied in several naturally existing microbial communities such as biofilms, where the geometry of the biofilm is also taken into account. In such cases, the spatial distribution of the organisms in the biofilm is modelled as a PDE. A more detailed explanation of PDE-based modelling can be found in the following section. For an elaborate discussion on the mathematical modelling of QS principles, we invite the reader to consult ref. [88].

3.1.2 PDE-based modelling

Many naturally occurring microbial communities display complex spatial relationships. As described in Chapter 1 (§1.2.3), such complex spatial relationships are well elucidated in biofilms, where each layer hosts a specific set of micro-organisms, depending on the spatial availability of resources. In these cases, the dynamics of the system are modelled using PDEs to determine the population dynamics.

The most commonly used PDE-based model is based on the reaction–diffusion equation, which captures the population density of each species with respect to time and space [89, 90]. This can be expressed as follows:

$$\frac{\partial C_k}{\partial t} = D_k \left(\frac{\partial C_k}{\partial x} + \frac{\partial C_k}{\partial y} + \frac{\partial C_k}{\partial z} \right) + r_k(x, y, z, C_1, C_2, \ldots, C_k)$$

$$k = 1, 2, \ldots, K$$

(3.4)

where C_k is the population density of species k at a given time t

D_k is the diffusion coefficient of species k

x, y, z are the coordinates in space

r_k is the growth rate or the death rate of the kth species in the microbial community.

This generalised PDE-based model has been suitably modified to study different types of ecological phenomena related to capturing spatial relationships [58, 91–93]. In one such study [58], the range expansion of an ecological community was modelled, and it was found that co-operators increased as the range expanded. In another related study [92], the range expansion was modelled from an experiment involving a mixed population of co-operating and cheating yeast strains. Here again, it was observed that co-operation is enhanced in expanding populations since the number of cheater strains was considerably reduced.

PDEs have also been used in conjunction with QS to model cell–cell interactions in biofilms. Specifically, the spatio-temporal dynamics of the biofilm formed by *P. aeruginosa* was modelled using PDEs to derive the relationships between the physical parameters and signalling molecules [94]. This study suggested that the organisms near the oxygen-deficit regions of biofilms have to continuously synthesise the signalling molecules. Lastly, PDEs have also been used along with constraint-based modelling techniques; more details can be found in §4.6.

3.1.3 Game-theoretic models

Game-theoretic models are mathematical frameworks to evaluate the outcomes of strategic interactions in a multi-organism population, where the pay-off depends not only on the strategies adopted by one organism but also on the collective action of all the organisms in the community. This pay-off is generally represented as a matrix and is used to calculate the equilibrium state of the population.

The concepts from game theory have also been extended to study population evolution. This branch of study is commonly known as *evolutionary game theory*. Here, the pay-off values in the matrix are used to describe the reproductive fitness of the organisms. The reproduction dynamics of interacting populations have been modelled using the following simple replicator's equation [95,96]:

$$\frac{dx_k}{dt} = (f_k(\mathbf{x}) - \phi(\mathbf{x}))x_k \tag{3.5}$$

$$k = 1, 2, \dots, K$$

$$f_k(\mathbf{x}) = \sum_{k'=1}^{K} a_{kk'} x_{k'} \tag{3.6}$$

$$k = 1, 2, \dots, K$$

$$\phi(\mathbf{x}) = \sum_{k'=1}^{K} f_k'(\mathbf{x}) x_k' \tag{3.7}$$

where $\mathbf{x} = [x_1, x_2, \dots, x_K]^T$, x_K represents the relative abundance of species K

$f_k(\mathbf{x})$ represents the fitness of species k

$\phi(\mathbf{x})$ denotes the average fitness of the community

$a_{kk'}$ is the fitness (pay-off value) of organism k while in the presence of k'.

It is important to note that this model does not account for the changes in interactions due to mutations. Moreover, in many cases, it might be difficult to quantify the pay-off values used in the model equations. Nevertheless, game-theory based models have emerged as an important mathematical tool to study microbial interactions.

In one such example [97], the famous game-theoretic principle called the *snowdrift game* [98] was experimentally demonstrated using two types of *Saccharomyces cerevisiae* strains. The choice of the player's strategy in this game depends on the actions of the opponent, i.e., if the opponent cheats, the player should yield and vice-versa. *S. cerevisiae* produces the enzyme invertase when grown on sucrose. The monosaccharides generated from sucrose hydrolysis diffuse away into the surroundings. This wild-type strain was cultivated along with a mutant strain that does not produce invertase. Both types of strategies, i.e., wild-type outnumbering the cheaters and vice-versa, were observed under a wide range of experimental conditions. A game-theoretic model describing these interactions was also constructed, where the fitness function defined by equation 3.7 was suitably modified to include the experimental data. This model could also explain the strategy of the snowdrift game between the cheaters and the co-operators.

In addition, several experiments [99–101] indicate that the ecological strategies are highly dependent on geometry and spatial diffusions. Such information has also been incorporated in the game-theoretical models to capture the dependence of spatial structure and the strategies adopted.

Game-theoretic models have also been used to analyse the suitability of microbial consortia for biotechnological applications. For instance, biotechnological parameters like growth rate and substrate uptake rate were used in developing a game-theoretical model to describe the population dynamics and extracellular enzyme synthesis [102]. This model could predict the existence of three strategies depending on the metabolic cost involved in the production of exo-enzymes, namely co-operation, coexistence and competition. These strategies allude to the fundamental concepts of game theory, namely harmony game, snowdrift game and Prisoner's Dilemma respectively.

3.2 AGENT-BASED MODELLING

Microbial ecosystems can also be modelled by taking into account the characteristics of individual organisms in a population, such as the growth rate of the individual organism, specific substrate uptake rate and interactions between every individual in the population and other individuals. The properties of a given population/ecosystem result from the *collective behaviour* of the individual members of a community. This type of modelling is widely referred to as *agent-based modelling* or *individual-based modelling (IbM)*.

Several experiments have revealed the heterogeneity of individual members [103–105] in a population. Further, the local interactions between the cells assume a lot of importance as they shape the structure of microbial communities. These interactions are predominant, especially in spatially structured microbial communities.

IbMs have been used extensively to capture such changes observed due to the variations in individual cells. Early studies pertaining to IbMs have used the properties of the bacterium to model the colony growth. BacSim [106] is one such study that seeks to model the behaviour of bacteria taking into account different bacterial properties such as substrate uptake rate, cell death and cell division. Recently, an updated version of BacSim, named BSim 2.0 was released [57], which includes several new additional features, such as delayed differential equations (DDEs) to understand the intracellular dynamics and a 3D cell morphology.

In another study [107], iDynoMiCS (individual-based Dynamics of Microbial Communities Simulator) was developed, which seeks to understand the individual dynamics in a well-mixed reactor setup, namely a chemostat. iDynoMiCS, like the previous approaches, uses a smaller representative volume, which is then scaled up depending on the size of the reactor. This technique makes an important assumption that the reactors are well-mixed and that the whole reactor consists of replicates of the representative volume.

A more generalised framework based on IbM was recently proposed to model multi-species interactions in biofilms or cell flocs [108] on a three-dimensional scale. Here, the biological systems are modelled as cuboids, where the micro-organisms such as those found on biofilms adhere to the surface of these cuboids. The model consists of several components denoted as *sub-models*, each describing a particular concept, namely, emergence, sensing, interactions and stochasticity. Each of these sub-models comprises specific sets of equations describing the physical and biological processes. The applications of this framework were demonstrated for predicting the surface morphology of biofilms. In addition, this model can also be used to understand the emergent properties of whole microbial systems.

Key points

- Microbial community structure is defined by interactions both at the population level and individual level

- Population and spatial dynamics of microbes in a community can be modelled using ODEs and PDEs

- Strategic relationships between the micro-organisms can be well understood using game-theoretical models

- Individual-based or agent-based models capture the collective behaviour of a community, arising out of individual behaviours and characteristics

FURTHER READING

- Succurro A and Ebenhöh O (2018) Review and perspective on mathematical modeling of microbial ecosystems. *Biochem Soc Trans* **46**(2):403–412

- Cavaliere M, Feng S, Soyer OS, and Jiménez JI (2017) Cooperation in microbial communities and their biotechnological applications. *Environ Microbiol* **19**(8):2949–2963

CHAPTER 4

Constraint-based modelling of microbial communities

Constraint-based modelling comprises a collection of techniques, which have been commonly used to predict the *metabolic phenotype* of organisms at steady state. Organisms in nature confront different types of environments, harbouring different sets of resources. In order to survive, the organisms adapt by fine-tuning themselves to the exposed environments. Often, they tend to rewire their metabolism in response to the dynamic changes in the environment. The final phenotype must satisfy environmental constraints as well as other endogenous cellular constraints.

The most commonly used constraint-based analysis technique is Flux Balance Analysis (FBA) [109–111], which mathematically analyses the metabolite flow through an organism's reaction network. This technique operates on genome-scale metabolic *reconstructions*, which catalogue all known reactions happening inside the cells. FBA has been successfully demonstrated in individual micro-organisms to answer a wide variety of questions ranging from identifying targets for metabolic engineering to predicting lethal phenotypes for drug target identification [112].

FBA has also been modified to account for dynamic changes

in metabolite concentrations. Such a formulation is referred to as *dynamic FBA (dFBA)*, which seeks to extend FBA to account for dynamic changes. FBA has also been extended to microbial communities, where *community genome-scale models* of microbes have been used to predict metabolic interactions between the organisms, as we will see later in this chapter.

In the following section, we introduce the basic concepts of FBA as well as how genome-scale metabolic networks are reconstructed. In the remaining sections, we describe various constraint-based methods that have been developed to study microbial communities and finally conclude with a section on case studies.

4.1 FUNDAMENTALS OF CONSTRAINT-BASED MODELLING

In this section, we outline the fundamental aspects of constraint-based modelling and give an overview of the most commonly used technique, FBA. For a more detailed account of FBA and related techniques, the reader is referred to the classic textbook on the topic [113].

4.1.1 Flux Balance Analysis (FBA) − at a glance

FBA is a mathematical technique used to identify the flow or flux of metabolites through different reactions in a metabolic network. FBA involves solving a linear system of equations, which are derived from the *stoichiometric matrix* $S_{m \times r}$ consisting of the stoichiometric coefficients of metabolites that participate in every reaction. Every column of S represents a reaction, while every row represents a metabolite. The product of the stoichiometric matrix S and the flux vector v, represents the time derivatives of the concentration vector x, i.e.,

$$\frac{d x}{d t} = S \cdot v \tag{4.1}$$

Under steady state conditions, where there is no accumulation of metabolites,

$$\frac{d x}{d t} = S \cdot v = 0 \tag{4.2}$$

Two types of constraints are typically imposed over each reaction:

- *Thermodynamic* constraints, which determine if the reactions can be reversible or not

- *Capacity* constraints, which restrict the possible range of fluxes that every reaction can take up

Under these constraints, FBA seeks to solve an objective function and identify the most probable network configurations and flux distributions that can be taken by the system under consideration. Very often, this objective function is set to maximisation of growth rate, since it is assumed that the organisms tend to maximise their growth in rich nutrient conditions. However, the choice of objective function is often dictated by the rationale or the application, e.g., maximisation of ATP production, minimisation of nutrient consumption or maximisation of product formation [114]. Once the objective function and the constraints are set, the system is then solved for determining the steady state flux distributions, which indicate the metabolic capabilities of the organism.

Mathematically, FBA can be stated as

$$\max_{v} \ \mathbf{c}^{T}\mathbf{v}$$
$$\text{s. t.} \ \ \mathbf{S} \cdot \mathbf{v} = 0 \tag{4.3}$$
$$LB_r \leqslant v_r \leqslant UB_r \quad \forall r \in R$$

where
$\mathbf{c}$ is the (linear) objective function
$\mathbf{v}$ is the vector of reaction fluxes (the *flux distribution*)
$\mathbf{S}$ is the stoichiometric matrix of the system, with m rows and r columns
LB and UB are the lower and upper bounds of the reaction fluxes
R is the set of reactions in the model.

FBA: Key terms

Stoichiometric matrix This $m \times r$ matrix connects the rate of change of metabolite concentrations to the flux of each reaction, as described in Eq. 4.1. Normally, $r > m$; therefore, there are more unknowns (the r reaction fluxes) than equations (m metabolites in steady state), i.e., the system (4.3) is under-determined.

Constraints Since the under-determined system has infinitely many solutions, the feasible solution space is constrained by using capacity and thermodynamic constraints.

Objective function Despite the addition of constraints, the system is often still under-determined, and one way to identify a possible solution is by optimising an objective function, which is meant to capture the cell's goal in a given medium, such as maximum growth or minimum ATP consumption. Although many objectives are possible, the most commonly used objective function is that of biomass maximisation.

Biomass objective function The biomass objective function is normally a *fictitious* reaction that combines various metabolites in the cell in a net reaction that involves all important cellular components such as amino acids, cell wall components, lipids, co-factors and nucleotides.

Linear programming If the constraints are linear and the objective function is also linear, as in the case of the standard FBA, the formulation is a linear programming problem.

4.1.2   Reconstruction of genome-scale metabolic models

Genome-Scale Metabolic Models (GSMMs) consist of several components, which provide a holistic view of the metabolism inside any organism. The process of *reconstructing* a GSMM involves extracting information from the sequenced genome and annotating its contents by mapping onto different databases. Broadly, a GSMM comprises metabolites, reactions, genes and the Gene–Protein–Reaction (GPR) relationships. The reactions in a model can either describe intracellular metabolism, transport of metabolites between the compartments or exchange of metabolites with the surroundings. It is important that these reactions are mass and charge balanced.

Further, these reactions are often governed by a set of Boolean rules, denoted as GPR associations, which are derived from the relationships between genes and proteins. GPRs are often of the form (gene A or gene B), (gene A and gene B or gene C), depending on the association of genes to produce proteins catalysing a particular reaction. In addition, every reaction is assigned with appropriate lower and upper bounds depending on its reversibility and the range of fluxes it can carry. By default, the lower and upper bounds of reversible reactions are constrained to take values between $-\infty$ to ∞, while that of irreversible reactions are set to range between 0 and ∞. In practice, a value such as ± 1000 mmol gDW^{-1} h^{-1} is used instead of $\pm\infty$.

Further, every GSMM has a biomass reaction, which serves to quantify the growth rate of the organism, and is invariably used as the objective function. More details on the biomass objective function can be found elsewhere [115]. A detailed account of how to reconstruct metabolic networks is available in ref. [116]. GSMMs so reconstructed need to adhere to various standards, so that they can be readily exploited for simulation and inference [117], especially for modelling microbial communities.

Genome-scale Reconstruction: Some Useful Databases

ModelSEED [118] A very popular resource for reconstructing GSMMs and performing constraint-based modelling using these models.

MetaCyc and BioCyc [119] Multi-tier databases hosting a large collection of over 12,000 draft metabolic networks. It also provides tools for analysing and visualising metabolic networks.

KEGG [120] The classic database with over 10,000 reactions across thousands of organisms.

BiGG [121] BiGG consists of well-curated GSMMs of over 50 organisms.

Virtual Metabolic Human [23] An excellent resource for studying human–gut microbiome interactions, along with the AGORA database.

Path2Models [122] This database hosts the draft metabolic reconstructions built through automatic pipelines.

4.2 OPTCOM: A MULTI-LEVEL OPTIMISATION FRAMEWORK TO MODEL MICROBIAL CONSORTIA

OptCom [123] is one of the earliest methods developed to understand metabolism in microbial communities; it relies on a multi-level and multi-objective formulation. OptCom seeks to maximise the *community fitness*, without compromising on the fitness of individual organisms.

The problem formulation of OptCom consists of two levels: the inner level (problem), which specifies the fitness of individual organisms in the community, which is integrated to the outer level (problem) that optimises a community-level objective function. Such a formulation of OptCom ensures that the combined biomass of the community and the individual organisms is maximum. Further, OptCom has also been modified to address the cases where the organisms do not grow at their *maximum* growth rate, but exhibit co-operative behaviours. In such cases, more constraints are added to the inner problem, which specify that the biomass of the individual organisms can take a value within an optimal range ($vopt_{biomass}^{k}$). This optimal value of the biomass is calculated from the community perspective. This modified version, called *Descriptive OptCom*, also includes experimental data as constraints to the inner problem. OptCom has been successfully demonstrated for capturing both positive and negative types of interactions between the organisms in a community.

OptCom has also been extended to account for capturing the temporal dynamics of the microbial communities. This technique, called d-OptCom [124], integrates dynamic mass balance equations pertaining to the production of metabolites, the growth of organisms, and substrate uptake kinetics, to assess the biomass and the metabolites shared between the organisms.

4.3 COMMUNITY FBA

The primary focus of community FBA (cFBA) [125] is to determine microbial interactions and elucidate the metabolic capabilities of a community. cFBA represents a direct extension of FBA for a single organism. This is based on an important assumption of balanced growth and thus necessitates the attainment of metabolic steady state by microbial ecosystems. Given these conditions, cFBA seeks to identify the relative species abundances, flux distributions and the exchange fluxes of different metabolites.

The input to cFBA is the stoichiometric matrix of the microbial consortium, which can be readily constructed by merging the corresponding matrices from the constituent organisms. This matrix consists of the internal and transport reactions of individual organisms, unique extracellular reactions and cross-feeding reactions, which are derived based on the common set of exchange metabolites. The stoichiometric matrix of the consortium can be viewed as a *compartmentalised model*, where the organisms communicate with each other through a common extracellular medium.

To perform FBA at a community level, mass balance equations around every metabolite are first written, taking into account its production or consumption from all the organisms in a community, in addition to the exchange that happens between the microorganisms. More specifically, this equation consists of three major terms, corresponding to the metabolites produced or consumed through intracellular enzymatic reactions, biomass formation reaction and the reactions pertaining to the metabolic exchanges. The equation is mathematically written as

$$
\frac{dm_i}{dt} = \sum_{j=1}^{n_X} \sum_{k=1}^{n_R} n_{ik} q_{kj} X_j + \sum_{j=1}^{n_X} g_{ij} \mu_j X_j + \sum_{l=1}^{n_E} n_{il} J_l \quad (4.4)
$$

where m_i is the amount of ith metabolite
n_X is the number of microbial species in the community
n_R is the number of reactions producing the ith metabolite
n_E is the number of exchange reactions
n_{ik} is the stoichiometric coefficient of ith metabolite in reaction k
q_{kj} is the specific rate of reaction k in organism j
X_j is the total biomass
g_{ij} is the stoichiometric coefficient of metabolite m_i in the biomass reaction of organism j
μ_j is the specific growth rate of organism j
n_{il} is the stoichiometric coefficient of ith metabolite in reaction l
J_l is the rate of inflow or outflow of metabolite.

Since the community is assumed to be at metabolic steady state, the mass balance around the metabolite is assumed to be zero. This can happen in different scenarios; however, the primary focus of this method has been to understand the microbial community at balanced growth, in which case the specific growth rate of the microbial community need not be equal to the maximum growth rate of the constituent organisms. With respect to the total biomass of the community at any time t,

$$0 = \sum_{j=1}^{n_X} \sum_{k=1}^{n_R} n_{ik} q_{kj} \frac{X_j(t)}{\sum\limits_{j=1}^{n_X} X_j(t)} + \sum_{j=1}^{n_X} g_{ij} \mu_C \frac{X_j(t)}{\sum\limits_{j=1}^{n_X} X_j(t)} + \sum_{l=1}^{n_E} \frac{n_{il} J_{il}(t)}{\sum\limits_{j=1}^{n_X} X_j(t)} \tag{4.5}$$

Since this relationship holds true over the entire exponential phase, the biomass fraction, reaction rates and the growth rates are independent of time. Thus, the biomass fraction, a constant, is defined as follows:

$$\phi_j = \frac{X_j}{\sum\limits_{j=1}^{n_X} X_j(t)} \tag{4.6}$$

However, the problem definition of cFBA becomes non-linear if the biomass fraction ϕ is considered as a variable. The non-linearity is resolved assuming ϕ to be a parameter, and considering multiple values of ϕ. cFBA thus seeks to solve the following equation:

$$\max_q \mu_C$$

$$\text{s. t.} \quad \sum_{j=1}^{n_X} \phi_j \left(\sum_{k=1}^{n_R} n_{ik} q_{kj} + g_{ij} \mu_C \right) + \sum_{l=1}^{n_E} n_{il} q_{il} = 0 \tag{4.7}$$

$$q_{min} \leqslant q \leqslant q_{max} \quad \text{(capacity constraints)}$$

4.4 SMETANA: MULTILEVEL FRAMEWORK TO MAP METABOLIC INTERACTIONS

In order to evaluate the metabolic resource overlap and the interaction potential between the organisms in a community, SMETANA (species metabolic interaction analysis) was developed [126]. The method tries to identify all possible inter-species interactions under a given nutritional environment, in terms of three separate scores:

1. Species Coupling Score (SCS)

2. Metabolite Uptake Score (MUS)

3. Metabolite Production Score (MPS)

The first two scores are obtained using a Mixed Integer Linear Programming (MILP) based algorithm, while the third one is obtained using a Linear Programming (LP) based algorithm. The objective function for each algorithm varies, depending on the values to be calculated. SMETANA score is the sum of all inter-species dependencies, i.e., SCS, MUS and MPS, in a given environment. Below, we outline each of the algorithms.

Algorithm to calculate SCS

SCS identifies the dependencies of a given species A on another species B, when present together in a community. To this end, an MILP problem is formulated and solved to determine the minimal set of members (denoted as donating species) necessary to support the growth of A. The resultant set is then added as a constraint to the outer MILP problem, and this process is repeated till all such sets have been identified. SCS is calculated from each of the obtained sets, by calculating the fraction of solutions where species B appears. Mathematically, the SCS algorithm can be described as:

$$\min \sum_{s \in C \setminus A} \theta_s$$

$$\text{s. t.} \quad \mathbf{S_s} \cdot \mathbf{v_s} = 0, \forall s \in C$$

$$v_i^{lb} \leqslant v_i \leqslant v_i^{ub} \quad \forall i \in R_s$$

$$v_{A, \text{growth}} = 1$$

$$\sum_{s \in L} v_{s, \text{secretion}} - \gamma \times \theta_s \leqslant 0, \forall s \in C, \theta_s \in \{0, 1\}, \gamma > \text{argmax}(v)$$

$$v_{A, \text{growth}} - v_{A, \text{min\_growth}} \times \theta_s \geqslant 0, \forall s \in C \setminus A$$

$$\sum_{s \in L} \theta_s < |L|, \forall L \in \{\text{solutions so far determined}\}$$

$$-\epsilon \leqslant v_{\text{vitamin uptake}} \times (1 - v_{\text{measured vitamin uptake}}) \leqslant \epsilon$$

$$(4.8)$$

where θ_s is the binary constraint, which denotes the presence or absence of a member species secreting metabolites

$\mathbf{S_s}, \mathbf{v_s}$ represent the stoichiometric matrix and the flux distribution vector of a species s in a community C

v_i^{lb} and v_i^{ub} denote the lower and upper bounds for the fluxes of the reaction i in species s (R_s denotes the reactions in species s)

$v_{A, \text{growth}}$ is set to one to ensure the growth of species A

ϵ provides the bounds on vitamin uptake.

The last constraint represents the restriction on the uptake of vitamins, since it is assumed that the micro-organisms do not use it as their carbon source. The binary variable θ_s depicts the contribution of each member species on the community, i.e., if $\theta_s = 0$, the organisms do not secrete any metabolite.

Algorithm to calculate MUS

MUS measures the association of metabolites produced by the microbial community with the growth of individual species A. MUS is also calculated using an MILP algorithm similar to that used for calculations of SCS; however, the objective function is set to minimisation of the metabolite set donated to species A. The mathematical representation of MUS is given below:

$$\min \sum_{m \in \text{metabolites from A}} \theta_m$$

$$\text{s. t.} \quad \mathbf{S_s} \cdot \mathbf{v_s} = 0, \forall s \in C$$

$$v_i^{lb} \leqslant v_i \leqslant v_i^{ub} \; \forall i \in R_s$$

$$v_{A, \text{growth}} = 1 \tag{4.9}$$

$$v_m - \gamma \times \theta_m \leqslant 0, \forall m \in \{\text{metabolite uptakes from A}\}$$

$$\sum_{m \in L} \theta_m < |L|, \forall L \in \{\text{solutions so far determined}\}$$

$$-\epsilon \leqslant v_{\text{vitamin uptake}} \times (1 - v_{\text{measured vitamin uptake}}) \leqslant \epsilon$$

where θ_m is the binary constraint, which denotes the uptake or secretion of metabolites secreted by member species. Other variables follow the same conventions as in Equation 4.8.

Algorithm to calculate MPS

MPS is a binary score that determines whether a metabolite can be produced by a given species in a community. This can be readily obtained by using FBA, maximising for the metabolite production, with an additional constraint on vitamin uptake.

$$\max v_m$$
$$\text{s. t.} \quad \mathbf{S_s} \cdot \mathbf{v_s} = 0, \forall s \in C$$
$$v^{lb} \leqslant v \leqslant v^{ub} \tag{4.10}$$
$$-\epsilon \leqslant v_{\text{vitamin uptake}} \times \left(1 - v_{\text{measured vitamin uptake}}\right) \leqslant \epsilon$$

where v_m is the flux through the reaction producing metabolite m. Other variables follow the same conventions as in Equation 4.8.

4.5   DYNAMIC MODELLING OF CO-CULTURES

FBA, which operates at steady state, has also been extended to account for the dynamic changes in the concentration of metabolites by employing a set of differential equations. These equations describe the rate of substrate uptake or product formation, taking into account the kinetic parameters of the transporters involved. This approach, called dFBA, has also been successfully demonstrated for microbial mono-cultures [127], and more recently, even for co-cultures [128, 129].

Dynamic Multi-species Metabolic Modelling (DMMM) [129], based on dFBA, integrates kinetic equations to FBA, in a community framework. The growth of individual organisms in the microbial community is calculated as:

$$\frac{dX_j}{dt} = \mu_j X_j \tag{4.11}$$

where
X_j is the biomass concentration of species j
μ_j is the growth rate of species j.
Further, the rate of production or consumption of metabolites by micro-organisms in a community is influenced by the other organisms. This is given by:

$$\frac{dS_i}{dt} = \sum_{j=1}^{N} V_i^j X_j \tag{4.12}$$

where
S_i is the concentration of substrate or metabolite i in the environment

V_i^j is the specific rate of production or consumption of substrate or metabolite i by species j

X_j is the biomass of species j

N is the maximum number of species in the community.

The specific rate of production or consumption of substrate i by species j is then calculated using FBA, as explained in §4.1.1. The capacity constraints of reactions corresponding to the uptake or production of extracellular metabolites for FBA are calculated either from the extracellular concentrations or based on Michaelis-Menten kinetics. By integrating the concentration values of metabolites and biomass over multiple time steps, the temporal dynamics of microbial community can be easily understood.

4.6 COMETS: UNDERSTANDING SPATIO-TEMPORAL DYNAMICS OF MICROBIAL COMMUNITIES

Computation of Microbial Ecosystems in Time and Space (COMETS) [130] is a lattice-based algorithm based on dFBA, which seeks to model the spatio-temporal dynamics of microbial communities. Here, a physical 2D space is considered and discretised into a grid of size $(N \times M)$, with every location defined by a pair of coordinates (x, y), where $x = 1, \ldots, N$, $y = 1, \ldots, M$. Every box in the grid is considered a square of size $L \times L$, which consists of different microbial species. The biomass of species α, the amount and concentration of metabolite m present at the coordinates (x, y) are denoted as $B_{x,y}^{\alpha}$, $Q_{x,y}^{m}$ and $C_{x,y}^{m}$, respectively. Since this is a dynamic framework, the biomass and its associated variables change with time.

In order to capture the dynamics of microbial communities in a lattice, dFBA is carried out, where the uptake and the secretion fluxes are determined based on a concentration-dependent saturating function. This function is similar to Michaelis–Menten kinetics, and consists of saturation coefficients, describing the maximum amount of metabolite that can be taken up by the organism. These values are set as the upper bound of the respective reactions, and an FBA is carried out. At every time point, these values are updated in the box, based on discrete update rules:

$$B_{x,y}^{\alpha}(t + \Delta t) = B_{x,y}^{\alpha}(t) + B_{x,y}^{\alpha}(t) \cdot v_{growth}^{\alpha} \cdot \Delta t$$
$$Q_{x,y}^{m}(t + \Delta t) = Q_{x,y}^{m}(t) + B_{x,y}^{\alpha}(t) \cdot u_{m}^{\alpha} \cdot \Delta t$$

$$(4.13)$$

where v^{α}_{growth} is the growth rate of the species in (x,y), while u^{α}_m is the uptake or secretion rates of metabolite m.

To account for the metabolite diffusion across the lattice, standard 2D lattice equations are solved, with varying diffusion coefficients for different components. Diffusion of organisms or metabolites depends on the respective concentrations, which are predicted using dFBA starting with user-defined initial conditions.

4.7 CASE STUDIES

4.7.1 Modelling mutualistic interactions between *D. vulgaris* and *M. maripaludis*

Methane production in different environments depends on the relationships between sulphate reducers and methanogens [6]. To provide deeper insights into the interactions between organisms and understand if there were any metabolite exchanges that played a role in the conservation of energy, metabolic models of the organisms *Desulfovibrio vulgaris* Hildenborough and *Methanococcus maripaludis* S2 were constructed. The individual models comprised aggregated reactions pertaining to central carbon metabolism and amino acid biosynthesis. The biomass reactions were obtained by merging the relevant precursors. Further, the stoichiometric coefficients of the reactants in the biomass equation were adapted from those of the *E. coli* model.

The community stoichiometric model was constructed by merging the models of organisms through a separate compartment, which acts as the medium for organisms to communicate with the extracellular environment. This additional compartment also facilitates the transfer of metabolites between the organisms. The final co-culture model, which consisted of 170 reactions and 147 metabolites was analysed using FBA by maximising for the objective function comprising the biomass of both the organisms. The joint model was further constrained with the experimentally determined flux values for lactate and hydrogen. To account for the experimental variations in these fluxes, three sets of simulations were performed, using the average flux values and constraining the lower or the upper bounds of the reactions, assuming a 95% confidence interval of the measurements.

The model predictions of an increase in carbon dioxide and biomass fluxes correlated well with those observed experimentally. Further, the joint models also predicted the exchange of both hydrogen and formate as the source of electron transfer, with hydrogen assuming primary importance. The flux distributions also suggested the dominant role of hydrogen transfer during the growth on lactate medium.

4.7.2 Dynamic modelling of microbial co-cultures

To identify the combination of organisms best suited for degrading the sugars in biomass feedstocks, a dFBA of microbial co-cultures was carried out [128]. Different microbial co-culture systems were modelled to examine the effect of multiple parameters on the substrate uptake rate and product formation rate. To this end, stoichiometric models of *Escherichia coli* iJR904 and *Saccharomyces cerevisiae* iND750 were used, with appropriate genetic modifications pertaining to carbon source utilisation. Further, the uptake of multiple substrates such as glucose, xylose, oxygen and ethanol was calculated using different sets of kinetic equations:

$$v_g = v_{g,max} \frac{G}{(K_g + G)} \frac{1}{(1 + \frac{E}{K_{ie}})}$$

$$v_z = v_{z,max} \frac{Z}{(K_z + Z)} \frac{1}{(1 + \frac{E}{K_{ie}})} \frac{1}{(1 + \frac{G}{K_{ig}})} \quad (4.14)$$

$$v_o = v_{o,max} \frac{O}{(K_o + O)}$$

where G, Z, O and E are the extracellular concentrations of glucose, xylose, oxygen and ethanol respectively, and $v_{g,max}$, $v_{z,max}$, $v_{o,max}$, K_g, K_z, K_o, K_{ie} and K_{ig} represent the maximum uptake rates, saturation and inhibition constants corresponding to these substrates and products. The values for these parameters were obtained from the literature. The co-culture is assumed to consist of organisms that are non-interacting, and hence the FBA was carried out using the following objective function:

$$\max_{v_i, v_j} \mu = \mu_i + \mu_j = w_i^T v_i + w_j^T v_j$$

$$\text{s.t.} \begin{bmatrix} A_i & 0 \\ 0 & A_j \end{bmatrix} \begin{bmatrix} v_i \\ v_j \end{bmatrix} = \begin{bmatrix} 0 \\ 0 \end{bmatrix} \quad (4.15)$$

$$\begin{bmatrix} v_{i,min} \\ v_{j,min} \end{bmatrix} \leqslant \begin{bmatrix} v_i \\ v_j \end{bmatrix} \leqslant \begin{bmatrix} v_{i,max} \\ v_{j,max} \end{bmatrix}$$

Using this paradigm, optimal ratios of the organisms in co-culture, substrate uptake parameters and the co-culture growth conditions were identified.

In another such study [131], the GSMMs of *Saccharomyces cerevisiae* and *Pichia stipitis* were used to understand the behaviour of these organisms in co-culture. Kinetic equations pertaining to substrate uptake and product formation were used to perform a dFBA on the joint models built on the assumption that the organisms do not interact with each other. The dFBA was used to determine the optimal concentration of microbes in the inoculum, aeration levels and the maximum ethanol productivity.

4.7.3 Constraint-based modelling of a consortium producing vitamin C

Vitamin C is an industrially important chemical, primarily synthesised through two approaches [132]: *Reichstein process*, which is a combination of six chemical steps and a single fermentation step, and the *modern two-step fermentation*, which involves two chemical and fermentation steps each. The latter method accounts for more than 80% of the Vitamin C produced worldwide. In this process, the intermediate steps pertaining to the conversion of L-sorbose to 2-keto-L-gulonic acid (2-KLG) are carried out using a synthetic microbial consortium comprising *Ketogulonicigenium vulgare* and *Bacillus megaterium*.

Several genomic, proteomic and metabolomic studies [47, 133, 134] indicate that the interactions between *B. megaterium* and *K. vulgare* are either mutualism or amensalism. In addition, using the genome sequences, metabolic models of *K. vulgare* WSH001 and *B. megaterium* WSH002 have been constructed [135, 136]. Later, these models were merged to represent an *Artificial Metabolic Ecosystem* to investigate the metabolic relationships between these organisms [52]. The combined model predominantly had four compartments — one pertaining to each of the two organisms, and the media for communication with the environment and between the organisms. FBA of the individual microbes in minimal glucose medium indicated that *K. vulgare* does not grow individually. However, when grown together with *B. megaterium*, this co-culture exhibits growth, indicating a possible transfer of metabolites from *B. megaterium* to *K. vulgare*. Further, FBA also predicted the exchange of 24 other metabolites between these organisms.

Key points

- Constraint-based modelling is an effective and popular approach to predict the phenotype of single organisms, and can be well extended to study microbial communities

- Metabolic networks can be reconstructed from pathway databases, literature and are informed by genome sequence

- Methods such as OptCom and cFBA devise modified objective functions to capture interactions between organisms in communities

- dFBA-based methods have been used to capture the dynamics of interactions between microbes in communities

- Applications demonstrated thus far include dynamic modelling of co-cultures and production of vitamin C

FURTHER READING

- Feist AM, Herrgård MJ, Thiele I, Reed JL, and Palsson BØ (2009) Reconstruction of biochemical networks in microorganisms. *Nat Rev Microbiol* **7**(2):129–143

- Palsson BØ (2015) *Systems biology: constraint-based reconstruction and analysis.* Cambridge University Press. ISBN 1107038855

- Maranas CD and Zomorrodi AR (2016) *Optimization Methods in Metabolic Networks.* Wiley, 1/e. ISBN 1119028493

- Magnúsdóttir S and Thiele I (2017) Modeling metabolism of the human gut microbiome. *Curr Opin Biotechnol* **51**:90–96

CHAPTER 5

Experimental techniques to understand microbial interactions

Computational techniques to model microbial communities have been extensively discussed in the previous chapters. Alongside the modelling tools, experimentally understanding the composition of a community and the interactions between the organisms plays an important role in discerning several important aspects of a community. These experiments can be broadly classified based on the type of data generated, into genomics, proteomics and metabolomics. Each of these techniques aids in understanding microbial communities through different lenses. Moreover, many algorithms and techniques build on these datasets to obtain deeper understanding and make useful predictions. In this chapter, we discuss a few techniques to discern the complex web of interactions happening in a microbial community. We broadly divide the contents of this chapter into two major categories, one pertaining to techniques for identifying and quantifying the micro-organisms in a community sample, and the other describing methods for identifying metabolic interactions happening between the micro-organisms.

5.1   IDENTIFICATION, DIFFERENTIATION AND QUANTIFICATION

Identification of individual organisms in a microbial community helps in deconvoluting and understanding the roles played by each organism. In addition, quantifying the individual organisms provides a better idea about the relative importance of each organism in the community. This also helps in ascertaining the functional roles of each organism and understanding the population dynamics. Such information can be used to answer questions such as, "How does the gut microbiome change when afflicted by a disease?" or "Does the microbiome in the soil vary depending on the region?"

The experimental techniques to study microbial communities can be broadly classified into microbiological and molecular biological. The former set of methods provides the general features of the community, while the latter set of methods aids in assigning the members of the consortium to individual taxa. In this section, we list a few of the most commonly used experimental techniques to characterise microbial communities.

5.1.1   Techniques based on molecular biology

Molecular biology-based techniques are used to quantify and identify the species in a microbial community. There exists a comprehensive suite of methods that is used for isolation of nucleic acids such as the DNA and the 16S-rRNA, which are then used to identify the constituent organisms. The nucleic acids so isolated are further sequenced to yield information about microbial populations in the community. Here, we describe some of the most commonly used molecular biology methods and discuss their pros and cons.

5.1.1.1   *Polymerase Chain Reaction*

Polymerase Chain Reaction (PCR) provides an elegant way to extract specific nucleic acid sequences by amplification from a large pool of DNA. The PCR reaction mixture consists of specific oligonucleotides (primers) and the enzyme DNA polymerase re-suspended in an appropriate buffer. During PCR, the complementary DNA strands break apart at high temperatures, followed by annealing of primers to specific regions of the DNA. Subsequently, DNA polymerase extends these primers, synthesising a new strand of DNA as it moves forward. The primers are designed such that they bind to template DNA and amplify a specific region of interest. This process is repeated for a defined number of cycles, thereby creating multiple copies of the region of interest.

In the case of microbial communities, the primers are designed such that they target the conserved regions such as the 16S-rRNA or 18S-rRNA sites. These regions serve as proxies to identify the taxonomy of micro-organisms forming a part of the community. The PCR products can then be sequenced to assign the organisms to a particular taxon.

Several experimental considerations have to be taken while using PCR-based techniques to analyse microbial communities. These include the choice of primers targeting the conserved sequences of the micro-organisms and the varying %G+C content of the nucleic acids of the microbial communities. Furthermore, extension by *Taq* polymerase can introduce a few errors, which would be a problem when sequenced.

5.1.1.2   Denaturing Gradient Gel Electrophoresis

Denaturing Gradient Gel Electrophoresis (DGGE) is a technique used to separate the DNA fragments depending on the mobility differences caused due to the differences in the %G+C content of the DNA sequences. For this purpose, the 16S-rDNA sequences amplified using PCR are run through a vertical polyacrylamide gel with differential gradient concentrations of denaturants such as formamide and urea. The strands of the DNA denature and separate into single strands under the influence of these denaturants. The relative proportions of G + C and A + T contents affect the denaturation of DNA, due to which they are separated at different positions within the gel. The amplicons with lower melting temperature (i.e., lower %G+C content) start denaturing earlier in the run, compared to those with high melting temperatures, thereby getting separated at different positions along the polyacrylamide gel.

DGGE is one of the most commonly used methods used to characterise microbial communities. The samples for DGGE are directly obtained from the PCR of microbial communities. However, depending on the composition of microbial communities, DGGE can produce complex electrophoretic profiles, which may be tough to analyse. Further, DGGE works best for shorter DNA sequences and requires large quantities of amplicons for effective separation.

5.1.1.3 In situ hybridisation

This technique involves the binding of short and labelled oligonucleotide sequences to the DNA isolated from microbial communities. These short *oligos*, referred to as probes, are often labelled with fluorochromes or enzyme-linked detection systems. These are most commonly used to identify the taxa of constituent microorganisms and quantify their relative abundances. The process of probe design starts with the *in silico* alignment of rDNA sequences from the organisms to identify the conserved regions, followed by their synthesis and labelling. The signals released by the binding of these probes to specific regions of DNA are then measured and correlated to the relative abundances of each taxon. These signals vary depending on the type of probe used. It is imperative to optimise the binding conditions and sensitivities of the probe to ensure that the probes bind correctly to the region of interest.

5.1.1.4 Quantitative PCR

The principle of probe hybridisation to DNA sequences has also been used in conjunction with PCR. This technique, commonly known as *quantitative PCR* (qPCR) or *real-time PCR* helps in monitoring the DNA amplification at every cycle of the PCR. Such real-time analysis is aided by the presence of fluorescent molecules such as fluorescent hybridisation probes, DNA-binding dyes and molecular beacons. These molecules fluoresce upon binding to the DNA; the fluorescence intensity is then measured and correlated to the amount of DNA present in the sample. The most widely used detection methods use SYBR Green I or TaqMan assays to quantify the micro-organisms present in a sample.

SYBR Green I is a dye that binds to the minor groove of DNA and emits fluorescent signals. During every cycle of PCR, as the specific regions of DNA are amplified, many molecules of SYBR Green I bind to the DNA, thereby leading to increased fluorescence. However, it is important to note that the binding of SYBR

Green I to DNA strands is non-specific, thereby leading to higher incidences of false positives. Due to its low cost and simplicity, SYBR Green I is most predominantly used in analysing microbial communities.

TaqMan assays involve TaqMan probes, whose 5′ end and 3′ end have a reporter dye and a quencher dye respectively. When both these ends are close to each other, the fluorescence emitted by the reporter dye is absorbed by the quencher dye through a phenomenon called fluorescence resonance energy transfer (FRET). Upon binding to the specific region of interest during PCR, these two dyes are split apart due to the exonuclease activity of DNA polymerase. Due to this, the fluorescence emitted by the reporter dye can be observed.

Quantification of microbial communities using these techniques involves identifying the threshold cycle value (Ct). Ct is defined as the number of cycles for the fluorescence to exceed the threshold levels within the exponential phase of the PCR. This implies that the higher the copies of target DNA sequence, the lower the Ct value, i.e., the number of cycles required for exhibiting higher fluorescence.

5.1.2   Techniques based on microbiology

Techniques under this category are broadly based on the physiological characteristics of micro-organisms such as the composition of the cell membrane and ability to catabolise multiple carbon sources. Depending on the technique, each of these provides different types of information. For instance, a simple cell counting technique gives an idea about the number of micro-organisms present in the community. On the other hand, methods such as flow cytometry and cell sorting allow the separation of micro-organisms from a community. In this section, we elaborate on a few microbiology-based techniques that are widely used in community analysis.

5.1.2.1   Cell counting techniques

Cell counting methods are widely used to identify the number of species present in a community. However, they do not provide any information about the phylogeny of the constituent organism. Moreover, these methods can be laborious and time-consuming.

There are two major categories, viz. direct cell counting and indirect cell counting.

Direct cell counting methods use different types of dyes to stain the DNA of microbial cells present in a community. These stained organisms are then visualised under the microscope and counted. The most widely used fluorescent dyes include Acridine Orange and 4′,6-diamidino-2-phenylindole (DAPI). These dyes, when bound to the DNA, emit green and blue colours, which can be observed at 525nm and 461nm respectively. This method also does not entail culturing the micro-organisms, due to which it is simple and easy to use.

Indirect cell counting methods, on the other hand, involve growing the micro-organisms on a complex medium and determining the number of colonies. Different measures such as Colony Forming Units (CFU) and Most Probable Number (MPN) exist, which help in estimating the number of micro-organisms in the sample. In order to estimate the CFUs, the micro-organisms are plated on a solid medium, incubated at a given temperature for a specific time period, and the colonies are noted.

In contrast, Most Probable Number (MPN) is calculated through a serial dilution technique, where different dilutions of micro-organisms are prepared from the original sample. These dilutions are then scored, depending on their responses such as the development of turbidity and excretion of specific metabolites, which are later correlated to the relative abundances of micro-organisms. This method is more often used to estimate the microbial populations in the cases where it is difficult to use the enumeration technique based on plating.

5.1.2.2 Flow cytometry

Flow cytometry is a high-throughput technique that can be used to generate the fingerprint of microbial communities. Here, the micro-organisms are forced through a small nozzle by a flowing liquid, which drives these cells through a laser light so that at any given time, only one cell passes through. The cells, when passed through the laser light, scatter light, which is then captured by the detectors. Detectors are placed across and in front of the laser beam, which measure the side scatter (SSC) and the forward scatter (FSC) respectively. SSC gives an idea about the granularity of the cells, while FSC provides information about cell size. Fluorescence detectors are also present, which capture the fluorescence emitted by the *labelled* organisms tagged with fluorescent proteins. Flow cytometry provides a lot of information such as the cell numbers and cell size distribution. Many a time, analysing such large quantum of data requires methods from statistics and multivariate data analyses to study the effect of multiple parameters.

Several quantitative approaches have been developed to handle the large quantum of data generated by flow cytometry, particularly with respect to microbial communities [137, 138]. These techniques are predominantly based on different types of statistical and machine learning-based algorithms. Flow cytometry has been extensively used to understand microbial communities, especially those from aquatic habitats [139–141].

5.2 UNDERSTANDING METABOLIC INTERACTIONS AND FLUX DISTRIBUTIONS IN MICROBIAL COMMUNITIES

Micro-organisms in a community are often found interacting with each other and tend to carry out complex functions that are otherwise impossible by single organisms. In many cases, it is quite interesting to understand and identify these interactions so that these can be leveraged when microbial communities are applied for biotechnological applications. However, experimentally determining these metabolic interactions in microbial communities is quite tricky, owing to the complexities in distinguishing the metabolites arising from different organisms. Further, for integrating the experimental data with computational modelling of microbial communities, it may often be useful to determine the intracellular fluxes in the organisms.

Conventionally, the intracellular flux distributions in single organisms are identified by growing organisms on labelled substrates such as ^{13}C-glucose and ^{13}C-methanol, and analysing the amino acids, which serve as proxies to understand the carbon flow through different metabolites. However, this method cannot be directly extended to microbial communities, where a group of organisms are involved, since it is not possible to distinguish the metabolites and assign their identity based on the organism that secreted it. To address this, indirect approaches targeting specific peptides in the organisms have been developed [142–144], a few of which have been extended to study microbial communities.

In one of the early studies [143], ^{13}C-flux distributions were identified from a sub-population of cells from a mixed environment by determining the ^{13}C-labelling patterns in amino acids of a reporter protein. Specifically, a plasmid harbouring genes encoding Green Fluorescence Protein (GFP) and Glutathione S-transferase (GST) (as a high-affinity purification tag) was cloned in two different strains of *Escherichia coli* K-12 MG1655, each having a deletion in phosphoglucose isomerase (pgi protein) and malate dehydrogenase (mdh protein) respectively. Each of these genetically modified *E. coli* cells were grown as a co-culture with wild-type *E. coli* strains in a medium containing U-^{13}C-glucose. The reporter protein was purified from the biomass pellets and subjected to Gas Chromatography–Mass Spectrometry (GC–MS) analysis to identify the labelling patterns of amino acids, which was used to infer the ^{13}C-label distributions in the genetically modified *E. coli* phenotypes. However, this method might be laborious and time-consuming, since it involves isolating proteins from multiple organisms and subsequently identifying labelling patterns.

To address these issues, a peptide-based labelling method was developed to determine the ^{13}C-label distributions in microbial communities [144]. In this technique, the labelling patterns of peptides from micro-organisms in the communities are observed. The peptide sequences of the organisms can be easily obtained through high-throughput proteomics techniques. In principle, this method adopts the same methodology as that of determining the labelling patterns from amino acids. Experimentally, the average amino acid label distribution of the whole microbial community is obtained, and the fluxes are mapped to individual organisms using a set of equations. It is important to note that the correspondence of the amino acid labels to the metabolic fluxes is non-linear. Thus, the equations to calculate the fluxes from the *Mass Distribution Vector* (MDV) are suitably modified to account for the specific characteristics of microbial communities. We invite the reader to consult ref. [144] for further details.

Key points

- *In vitro* experiments are very useful in understanding several aspects of microbial communities, such as the relative abundances and the microbial interactions

- Molecular biology-based techniques help quantify the relative abundances and determine the phylogeny of organisms in communities

- Metabolic interactions and the flux distributions in microbial communities have been widely determined using tracer experiments involving ^{13}C-labelled substrates

- The genomic, proteomic and metabolomic data generated through experimental techniques can be analysed through a repertoire of modelling techniques

---

FURTHER READING

- Spiegelman D, Whissell G, and Greer CW (2005) A survey of the methods for the charcterization of microbial consortia and communities *Can J Microbiol* **51**(5):335–386

- Ghosh A, Nilmeier J, Weaver D, Adams PD, Keasling JD, Mukhopadhyay A *et al.* (2014) A peptide-based method for ^{13}C metabolic flux analysis in microbial communities. *PLoS Comput Biol* **10**(9):e1003827

CHAPTER 6

Outlook

Microbial communities are ubiquitous and are known to play important roles in several natural processes ranging from the regulation of biogeochemical cycles to the digestion of complex polysaccharides in human beings. Due to their immense potential, microbial consortia have attracted a lot of interest, particularly for their use in biotechnology. The applications of microbial communities are manifold — well extending from producing important chemicals to breaking down recalcitrant substrates.

Recent advances in high-throughput sequencing technologies have generated a large quantum of data at multiple omics strata, namely transcriptomic, metabolomic and proteomic. To understand these datasets and gain useful insights, a wide repertoire of computational methods has been developed. Each of these methods is tailored towards datasets of different types. Broadly, we classify these methods into four major categories, namely, network-based, population-based, individual-based and constraint-based techniques. Figure 6.1 broadly summarises these techniques.

Network-based modelling techniques construct a graph/network of microbial communities using the information from metagenomic sequences or metabolic networks, to quantify the relative abundances of each species and identify the metabolic interactions between the organisms. The number of metagenomic datasets has mushroomed in recent years owing to the advancements in sequencing technologies, holding great promise for fascinating insights. However, on the flip side, it is quite challenging to *connect the dots* between the sequencing reads and assign them to their respective taxa. Further, the methods under this cat-

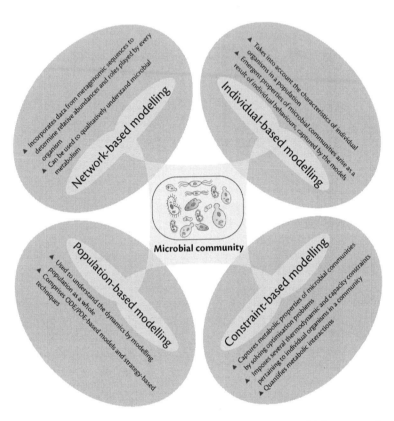

Figure 6.1 **Summary of modelling techniques:** Microbial communities can be broadly modelled using four different types of techniques. Network-based modelling techniques seek to understand the association and qualitatively decipher the interactions between the microbes in a community. Population-based and individual-based models capture the population and spatial dynamics at a collective and single organism level, respectively. Constraint-based modelling techniques have been used to predict the metabolic phenotype of microbial communities.

egory predominantly analyse static networks, while in many real-life scenarios, dynamics play an important role.

The other class of methods, falling under population-based and agent-based modelling techniques, model the spatial and population dynamics of microbial communities at collective and individual scales respectively. Such dynamics are commonly represented as ordinary and partial differential equations. In natural ecosystems, the dynamics of organisms is influenced by several parameters. However, solving a large model consisting of multiple equations and parameters is quite challenging.

Constraint-based methods, on the other hand, are useful in predicting the metabolic phenotype of microbial communities. These techniques help quantify the metabolic interactions between the organisms in a community, subject to specific sets of constraints. The resulting under-determined linear system of equations is solved under *steady state assumptions*, although there has been progress in incorporating dynamic data as well. It is important to note that these methods require well-curated metabolic networks and an accurate specification of *biologically realistic* objective functions. Currently, the number of such well-curated networks are very few in comparison to the number of draft reconstructions that have been generated using automatic pipelines. Nevertheless, constraint-based approaches have immense potential in quantifying the metabolic interactions in microbial communities.

Taken together, this comprehensive suite of modelling frameworks offers different perspectives to understand the underlying principles governing microbial communities. Community models have far-reaching applications, right from modelling better bioprocesses for production of industrially relevant chemicals to answering essential questions relevant to health and disease. These models can be further refined to improve predictions by integrating data from various types of experiments. Further advances in metagenomics, coupled with advances in meta-transcriptomics and metaproteomics, will present larger, more comprehensive and, at the same time, highly complex multi-dimensional datasets, which can further inform community modelling.

Besides, these modelling methods can be used to generate testable hypotheses. For instance, the pair-wise metabolic interactions predicted by network-based/constraint-based techniques can be validated using ^{13}C-labelling experiments. Similarly, the population dynamics as predicted by ODE-based models can be experimentally verified by quantitative PCR or fluorescent tagging. Such data can be used for iteratively refining the models and improving their predictions.

Although this book has focussed heavily on microbe–microbe interactions, the general principles of community modelling can be extended to model interactions between any cells. A few interesting studies have already explored this direction, beginning with the classic study on the interactions between *Mycobacterium tuberculosis* and human macrophage cells [145]. Another interesting study catalogued the interactions between *Plasmodium falciparum* and the erythrocyte, predicting metabolic exchanges between the host and pathogen [146]. Other studies have also modelled the interactions between phages and mycobacteria [147,148]. Computational approaches have also shed light on the interactions between the gut microbiota and the human, across different dietary regimes [149].

One of the biggest breakthroughs in computational systems biology has been the development of an accurate *whole-cell model* of *Mycoplasma genitalium* [150], and an increasing interest in building whole-cell models of humans [151]. The next step in the evolution of whole-cell models would be community models involving human and pathogen or two different microbes, to study and characterise the interactions at an unprecedented level of detail.

Another exciting modelling challenge is the construction of a multi-scale model of microbial communities. There are many complex microbial ecosystems (as detailed in §1.2) that can be modelled to answer many interesting questions. Beginning with models of how individual cells grow, and then embed in communities through metabolic exchanges and other interactions, to how microbial populations vary in an ecosystem, models are possible at every level of complexity. Such detailed multi-scale models can be a powerful tool to probe interactions between microbes in different ecosystems and ask important "what if" questions, such as "How does a perturbation of the gut microbiome affect its composition and ultimately human health?" or "How does a change in the composition of oceanic microbial community affect the biosphere?".

The last decade or so has seen many exciting advances in modelling microbial communities. In this book, we have attempted to capture the breadth of these methodologies. It is certain that the coming years will continue to build on these many advances towards more accurate models of microbial communities, microbe–microbe and pathogen–host interactions, across various spatio-temporal scales. Advances in experimental techniques that can accurately distinguish different players in communities, quantitate their metabolomes and fluxomes will likely have a key impact in the evolution of the modelling methodologies.

Bibliography

[1] Zuñiga C, Zaramela L, and Zengler K (2017) Elucidation of complexity and prediction of interactions in microbial communities. *Microb Biotechnol* **10**(6):1500–1522

[2] Kolenbrander PE, Palmer RJ, Periasamy S, and Jakubovics NS (2010) Oral multispecies biofilm development and the key role of cell-cell distance. *Nat Rev Microbiol* **8**(7):471–480

[3] Zhang H, Pereira B, Li Z, and Stephanopoulos G (2015) Engineering *Escherichia coli* coculture systems for the production of biochemical products. *Proc Natl Acad Sci U S A* **112**(27):8266–8271

[4] McInerney MJ, Struchtemeyer CG, Sieber J, Mouttaki H, Stams AJM, Schink B *et al.* (2008) Physiology, ecology, phylogeny, and genomics of microorganisms capable of syntrophic metabolism. *Ann N Y Acad Sci* **1125**:58–72

[5] Walker CB, He Z, Yang ZK, Ringbauer JA, He Q, Zhou J *et al.* (2009) The electron transfer system of syntrophically grown *Desulfovibrio vulgaris*. *J Bacteriol* **191**(18):5793–5801

[6] Stolyar S, Van Dien S, Hillesland KL, Pinel N, Lie TJ, Leigh JA *et al.* (2007) Metabolic modeling of a mutualistic microbial community. *Mol Syst Biol* **3**(92)

[7] Bryant MP, Wolin EA, Wolin MJ, and Wolfe RS (1967) Methanobacillus omelianskii, a symbiotic association of two species of bacteria. *Archiv für Mikrobiologie* **59**(1-3):20–31

[8] Smid EJ and Lacroix C (2013) Microbe-microbe interactions in mixed culture food fermentations. *Curr Opin Biotechnol* **24**(2):148–154

[9] Tlaskalová-Hogenová H, Tpánková R, Kozáková H, Hudcovic T, Vannucci L, Tuková L *et al.* (2011) The role of gut microbiota (commensal bacteria) and the mucosal barrier in the pathogenesis of inflammatory and autoimmune diseases and cancer: Contribution of germ-free and gnotobiotic animal models of human diseases. *Cell Mol Immunol* **8**(2):110–120

[10] Faust K and Raes J (2012) Microbial interactions: from networks to models. *Nat Rev Microbiol* **10**(8):538–550

[11] Lindgren SE and Dobrogosz WJ (1990) Antagonistic activities of lactic acid bacteria in food and feed fermentations. *FEMS Microbiol Rev* **7**(1-2):149–163

[12] Marcobal A and Sonnenburg JL (2012) Human milk oligosaccharide consumption by intestinal microbiota. *Clin Microbiol Infect* **18**(4):12–15

[13] Nazaries L, Pan Y, Bodrossy L, Baggs EM, Millard P, Murrell JC *et al.* (2013) Evidence of microbial regulation of biogeochemical cycles from a study on methane flux and land use change. *Appl Environ Microbiol* **79**(13):4031–4040

[14] Wang B, Yao M, Lv L, Ling Z, and Li L (2017) The human microbiota in health and disease. *Engineering* **3**(1):71–82

[15] Ussar S, Fujisaka S, and Kahn CR (2016) Interactions between host genetics and gut microbiome in diabetes and metabolic syndrome. *Mol Metab* **5**(9):795–803

[16] Gritz EC and Bhandari V (2015) The human neonatal gut microbiome: A brief review. *Front Pediatr* **3**

[17] O'Sullivan A, Farver M, and Smilowitz JT (2015) The Influence of Early Infant-Feeding Practices on the Intestinal Microbiome and Body Composition in Infants. *Nutr Metab Insights* **8**(S1):1–9

[18] Nicholson JK, Holmes E, Kinross J, Burcelin R, Gibson G, Jia W *et al.* (2012) Host-gut microbiota metabolic interactions. *Science* **336**(6086):1262–1267

[19] Ruppin H, Bar-Meir S, Soergel K, Wood C, and Schmitt MJ (1980) Absorption of short-chain fatty acids by the colon. *Gastroenterology* **78**:1500–7

[20] Jandhyala SM (2015) Role of the normal gut microbiota. *World J Gastroenterol* **21**(29):8787

[21] Levy R and Borenstein E (2013) Metabolic modeling of species interaction in the human microbiome elucidates community-level assembly rules. *Proc Natl Acad Sci U S A* **110**(31):12804–12809

[22] Zelezniak A, Andrejev S, Ponomarova O, Mende DR, Bork P, and Patil KR (2015) Metabolic dependencies drive species co-occurrence in diverse microbial communities. *Proc Natl Acad Sci U S A* **112**(20):6449–6454

[23] Magnúsdóttir S, Heinken A, Kutt L, Ravcheev DA, Bauer E, Noronha A *et al.* (2016) Generation of genome-scale metabolic reconstructions for 773 members of the human gut microbiota. *Nat Biotechnol* **35**(1):81–89

[24] Shoaie S, Karlsson F, Mardinoglu A, Nookaew I, Bordel S, and Nielsen J (2013) Understanding the interactions between bacteria in the human gut through metabolic modeling. *Sci Rep* **3**:2532

[25] van der Ark KCH, van Heck RGA, Martins Dos Santos VAP, Belzer C, and de Vos WM (2017) More than just a gut feeling: constraint-based genome-scale metabolic models for predicting functions of human intestinal microbes. *Microbiome* **5**(1):78

[26] Mooshammer M, Wanek W, Hämmerle I, Fuchslueger L, Hofhansl F, Knoltsch A *et al.* (2014) Adjustment of microbial nitrogen use efficiency to carbon:nitrogen imbalances regulates soil nitrogen cycling. *Nat Commun* **5**:3694

[27] Štursová M, Žifčáková L, Leigh MB, Burgess R, and Baldrian P (2012) Cellulose utilization in forest litter and soil: identification of bacterial and fungal decomposers. *FEMS Microbiol Ecol* **80**(3):735–746

[28] Fierer N and Jackson RB (2006) The diversity and biogeography of soil bacterial communities. *Proc Natl Acad Sci U S A* **103**(3):626–631

[29] Zarraonaindia I, Owens SM, Weisenhorn P, West K, Hampton-Marcell J, Lax S *et al.* (2015) The soil microbiome influences grapevine-associated microbiota. *MBio* 6(2):e02527–14

[30] Marques JM, da Silva TF, Vollu RE, Blank AF, Ding GC, Seldin L *et al.* (2014) Plant age and genotype affect the bacterial community composition in the tuber rhizosphere of field-grown sweet potato plants. *FEMS Microbiol Ecol* 88(2):424–435

[31] Costerton JW (1999) Bacterial biofilms: A common cause of persistent infections. *Science* 284(5418):1318–1322

[32] Palmer RJ, Gordon SM, Cisar JO, and Kolenbrander PE (2003) Coaggregation-mediated interactions of *Streptococci* and *Actinomyces* detected in initial human dental plaque. *J Bacteriol* 185(11):3400–3409

[33] Gilbert P, Das J, and Foley I (1997) Biofilm susceptibility to antimicrobials. *Adv Dent Res* 11(1):160–167

[34] Davey ME and O'Toole GA (2000) Microbial biofilms: from ecology to molecular genetics. *Microbiol Mol Biol Rev* 64(4):847–867

[35] Lee KWK, Periasamy S, Mukherjee M, Xie C, Kjelleberg S, and Rice SA (2014) Biofilm development and enhanced stress resistance of a model, mixed-species community biofilm. *ISME J* 8(4):894–907

[36] Li XZ, Webb JS, Kjelleberg S, and Rosche B (2006) Enhanced benzaldehyde tolerance in *Zymomonas mobilis* biofilms and the potential of biofilm applications in fine-chemical production. *Appl Environ Microbiol* 72(2):1639–1644

[37] Singh R, Paul D, and Jain RK (2006) Biofilms: implications in bioremediation. *Trends Microbiol* 14(9):389–397

[38] Zhang S, Norrlow O, Wawrzynczyk J, and Dey ES (2004) Poly(3-hydroxybutyrate) biosynthesis in the biofilm of *Alcaligenes eutrophus*, using glucose enzymatically released from pulp fiber sludge. *Appl Environ Microbiol* 70(11):6776–6782

[39] Wik T (2003) Trickling filters and biofilm reactor modelling. *Rev Environ Sci Biotechnology* **2**(2):193–212

[40] Jagmann N and Philipp B (2014) Design of synthetic microbial communities for biotechnological production processes. *J Biotechnol* **184C**:209–218

[41] Zuroff TR and Curtis WR (2012) Developing symbiotic consortia for lignocellulosic biofuel production. *Appl Microbiol Biotechnol* **93**:1423–1435

[42] Lynd LR, van Zyl WH, McBride JE, and Laser M (2005) Consolidated bioprocessing of cellulosic biomass: an update. *Curr Opin Biotechnol* **16**(5):577–583

[43] Lynd LR, Weimer PJ, van Zyl WH, and Pretorius IS (2002) Microbial cellulose utilization: Fundamentals and biotechnology. *Microbiol Mol Biol Rev* **66**(3):506–577

[44] Park EY, Naruse K, and Kato T (2012) One-pot bioethanol production from cellulose by co-culture of *Acremonium cellulolyticus* and *Saccharomyces cerevisiae*. *Biotechnol Biofuels* **5**:64

[45] Patle S and Lal B (2007) Ethanol production from hydrolysed agricultural wastes using mixed culture of *Zymomonas mobilis* and *Candida tropicalis*. *Biotechnol Lett* **29**:1839–1843

[46] Szambelan K, Nowak J, and Czarnecki Z (2004) Use of *Zymomonas mobilis* and *Saccharomyces cerevisiae* mixed with *Kluyveromyces fragilis* for improved ethanol production from Jerusalem artichoke tubers. *Biotechnol Lett* **26**:845–848

[47] Ma Q, Zhou J, Zhang W, Meng X, Sun J, and Yuan YJ (2011) Integrated proteomic and metabolomic analysis of an artificial microbial community for two-step production of vitamin C. *PLoS ONE* **6**(10):e26108

[48] Zhou J, Ma Q, Yi H, Wang L, Song H, and Yuan YJ (2011) Metabolome profiling reveals metabolic cooperation between *Bacillus megaterium* and *Ketogulonicigenium vulgare* during induced swarm motility. *Appl Environ Microbiol* **77**(19):7023–7030

[49] Saini M, Hong Chen M, Chiang CJ, and Chao YP (2015) Potential production platform of n-butanol in *Escherichia coli*. *Metab Eng* **27**:76–82

[50] Zhang H and Stephanopoulos G (2016) Co-culture engineering for microbial biosynthesis of 3-amino-benzoic acid in *Escherichia coli*. *Biotechnol J* **11**(7):981–987

[51] Nagarajan H, Embree M, Rotaru AE, Shrestha PM, Feist AM, Palsson BØ et al. (2013) Characterization and modelling of interspecies electron transfer mechanisms and microbial community dynamics of a syntrophic association. *Nat Commun* **4**:2809

[52] Ye C, Zou W, Xu N, and Liu L (2014) Metabolic model reconstruction and analysis of an artificial microbial ecosystem for vitamin C production. *J Biotechnol* **182-183**:61–67

[53] Tzamali E, Poirazi P, Tollis IG, and Reczko M (2011) A computational exploration of bacterial metabolic diversity identifying metabolic interactions and growth-efficient strain communities. *BMC Syst Biol* **5**:167

[54] Faust K, Lima-Mendez G, Lerat JS, Sathirapongsasuti JF, Knight R, Huttenhower C et al. (2015) Cross-biome comparison of microbial association networks. *Front Microbiol* **6**:1200

[55] Borenstein E and Feldman MW (2009) Topological signatures of species interactions in metabolic networks. *J Comput Biol* **16**(2):191–200

[56] Ravikrishnan A, Nasre M, and Raman K (2018) Enumerating all possible biosynthetic pathways in metabolic networks. *Sci Rep* **8**:9932

[57] Matyjaszkiewicz A, Fiore G, Annunziata F, Grierson CS, Savery NJ, Marucci L et al. (2017) BSim 2.0: An advanced agent-based cell simulator. *ACS Synth Biol* **6**:1969–1972

[58] Korolev KS (2013) The fate of cooperation during range expansions. *PLoS Comput Biol* **9**(3):e1002994

[59] Cormen TH, Stein C, Rivest RL, and Leiserson CE (2001) *Introduction to Algorithms*. McGraw-Hill Higher Education, 2nd/e. ISBN 0070131511

[60] Borenstein E, Kupiec M, Feldman MW, and Ruppin E (2008) Large-scale reconstruction and phylogenetic analysis of metabolic environments. *Proc Natl Acad Sci U S A* **105**(38):14482–14487

[61] Yassour M, Vatanen T, Siljander H, Hämäläinen AM, Härkönen T, Ryhänen SJ *et al.* (2016) Natural history of the infant gut microbiome and impact of antibiotic treatment on bacterial strain diversity and stability. *Sci Transl Med* 8(343):343ra81

[62] Faust K, Sathirapongsasuti JF, Izard J, Segata N, Gevers D, Raes J *et al.* (2012) Microbial co-occurrence relationships in the human microbiome. *PLoS Comput Biol* 8(7):e1002606

[63] Ponnusamy K, Choi JN, Kim J, Lee SY, and Lee CH (2011) Microbial community and metabolomic comparison of irritable bowel syndrome faeces. *J Med Microbiol* 60(6):817–827

[64] Li C, Lim KMK, Chng KR, and Nagarajan N (2016) Predicting microbial interactions through computational approaches. *Methods* 102:12–19

[65] Friedman J and Alm EJ (2012) Inferring correlation networks from genomic survey data. *PLoS Comput Biol* 8(9):e1002687

[66] Kurtz ZD, Müller CL, Miraldi ER, Littman DR, Blaser MJ, and Bonneau RA (2015) Sparse and compositionally robust inference of microbial ecological networks. *PLoS Comput Biol* 11(5):e1004226

[67] Xia LC, Ai D, Cram J, Fuhrman JA, and Sun F (2013) Efficient statistical significance approximation for local similarity analysis of high-throughput time series data. *Bioinformatics* 29(2):230–237

[68] Durno WE, Hanson NW, Konwar KM, and Hallam SJ (2013) Expanding the boundaries of local similarity analysis. *BMC Genomics* 14(Suppl 1):S3

[69] Shafiei M, Dunn KA, Boon E, MacDonald SM, Walsh DA, Gu H *et al.* (2015) BioMiCo: A supervised Bayesian model for inference of microbial community structure. *Microbiome* 3:8

[70] Handorf T, Ebenhöh O, and Heinrich R (2005) Expanding metabolic networks: Scopes of compounds, robustness, and evolution. *J Mol Evol* 61(4):498–512

[71] Christian N, Handorf T, and Ebenhöh O (2007) Metabolic synergy: increasing biosynthetic capabilities by network co-operation. *Genome Inform* **18**:320–9

[72] Freilich S, Kreimer A, Meilijson I, Gophna U, Sharan R, and Ruppin E (2010) The large-scale organization of the bacterial network of ecological co-occurrence interactions. *Nucleic Acids Res* **38**(12):3857–3868

[73] Levy R, Carr R, Kreimer A, Freilich S, and Borenstein E (2015) NetCooperate: a network-based tool for inferring host-microbe and microbe-microbe cooperation. *BMC Bioinformatics* **16**:164

[74] Carr R and Borenstein E (2012) NetSeed: a network-based reverse-ecology tool for calculating the metabolic interface of an organism with its environment. *Bioinformatics* **28**(5):734–735

[75] Kreimer A, Doron-Faigenboim A, Borenstein E, and Freilich S (2012) NetCmpt: a network-based tool for calculating the metabolic competition between bacterial species. *Bioinformatics* **28**(16):2195–2197

[76] Methé Ba, Nelson KE, Pop M, Creasy HH, Giglio MG, Huttenhower C *et al.* (2012) A framework for human microbiome research. *Nature* **486**(7402):215–221

[77] Huttenhower C, Gevers D, Knight R, Abubucker S, Badger JH, Chinwalla AT *et al.* (2012) Structure, function and diversity of the healthy human microbiome. *Nature* **486**(7402):207–214

[78] Arumugam M, Raes J, Pelletier E, Le Paslier D, Yamada T, Mende DR *et al.* (2013) Enterotypes of the human gut microbiome. *Nature* **473**(7346):174–180

[79] Naqvi A, Rangwala H, Keshavarzian A, and Gillevet P (2010) Network-based modeling of the human gut microbiome. *Chem Biodivers* **7**(5):1040–1050

[80] Cottret L, Milreu PV, Acuña V, Marchetti-Spaccamela A, Stougie L, Charles H *et al.* (2010) Graph-based analysis of

the metabolic exchanges between two co-resident intracellular symbionts, *Baumannia cicadellinicola* and *Sulcia muelleri*, with their insect host, *Homalodisca coagulata*. *PLoS Comput Biol* **6**(9):e1000904

[81] Steinway SN, Biggs MB, Loughran TP, Papin JA, and Albert R (2015) Inference of network dynamics and metabolic interactions in the gut microbiome. *PLoS Comput Biol* **11**(6):e1004338

[82] Murray JD (2002) *Mathematical biology*. Springer, New York. ISBN 9780387952239

[83] Stein RR, Bucci V, Toussaint NC, Buffie CG, Rätsch G, Pamer EG *et al.* (2013) Ecological Modeling from Time-Series Inference: Insight into Dynamics and Stability of Intestinal Microbiota. *PLoS Comput Biol* **9**(12):e1003388

[84] Fisher CK and Mehta P (2014) Identifying keystone species in the human gut microbiome from metagenomic timeseries using sparse linear regression. *PLoS ONE* **9**(7):e102451

[85] James S, Nilsson P, James G, Kjelleberg S, and Fagerström T (2000) Luminescence control in the marine bacterium *Vibrio fischeri*: An analysis of the dynamics of *lux* regulation. *J Mol Biol* **296**(4):1127–1137

[86] Dockery JD and Keener JP (2001) A mathematical model for quorum sensing in *Pseudomonas aeruginosa*. *Bull Math Biol* **63**:95

[87] Ward JP, King JR, Koerber AJ, Williams P, Croft JM, and Sockett RE (2001) Mathematical modelling of quorum sensing in bacteria. *IMA J Math Appl Med Biol* **18**:263–292

[88] Pérez-Velázquez J, Gölgeli M, and García-Contreras R (2016) Mathematical modelling of bacterial quorum sensing: A review. *Bull Math Biol* **78**(8):1585–1639

[89] Cantrell RS and Cosner C (2003) *Spatial ecology via reaction-diffusion equations*. John Wiley & Sons Ltd. ISBN 9780471493013

[90] Cosner C (2008) *Reaction–Diffusion Equations and Ecological Modeling*, pp. 77–115. Springer Berlin Heidelberg, Berlin, Heidelberg. ISBN 9783540743316

[91] Holmes EE, Lewis MA, Banks JE, and Veit RR (1994) Partial differential equations in ecology: Spatial interactions and population dynamics. *Ecology* **75**(1):17–29

[92] Datta MS, Korolev KS, Cvijovic I, Dudley C, and Gore J (2013) Range expansion promotes cooperation in an experimental microbial metapopulation. *Proc Natl Acad Sci U S A* **110**(18):7354–7359

[93] Müller MJI, Neugeboren BI, Nelson DR, and Murray AW (2014) Genetic drift opposes mutualism during spatial population expansion. *Proc Natl Acad Sci U S A* **111**(3):1037–1042

[94] Chopp DL, Kirisits MJ, Moran B, and Parsek MR (2002) A mathematical model of quorum sensing in a growing bacterial biofilm. *J Ind Microbiol Biotechnol* **29**(6):339–346

[95] Nowak MA (2006) *Evolutionary dynamics: exploring the equations of life*. Belknap Press. ISBN 9780674023383

[96] Zomorrodi AR and Segrè D (2016) Synthetic ecology of microbes: Mathematical models and applications. *J Mol Biol* **428**(5):837–861

[97] Gore J, Youk H, and Van Oudenaarden A (2009) Snowdrift game dynamics and facultative cheating in yeast. *Nature* **459**(7244):253–256

[98] Smith JM (1988) *Evolution and the Theory of Games*, pp. 202–215. Springer US, Boston, MA. ISBN 9781468478624

[99] Mark Welch JL, Hasegawa Y, McNulty NP, Gordon JI, and Borisy GG (2017) Spatial organization of a model 15-member human gut microbiota established in gnotobiotic mice. *Proc Natl Acad Sci U S A* **114**(43):E9105–E9114

[100] Stump SM, Johnson EC, and Klausmeier CA (2018) Local interactions and self-organized spatial patterns stabilize microbial cross-feeding against cheaters. *J R Soc Interface* **15**(140):20170822

[101] Goldschmidt F, Regoes RR, and Johnson DR (2018) Metabolite toxicity slows local diversity loss during expansion of a microbial cross-feeding community. *ISME J* **12**(1):136–144

[102] Schuster S, Kreft JU, Brenner N, Wessely F, Theißen G, Ruppin E *et al.* (2010) Cooperation and cheating in microbial exoenzyme production - Theoretical analysis for biotechnological applications. *Biotechnol J* **5**(7):751–758

[103] Brehm-Stecher BF and Johnson EA (2004) Single-cell microbiology: tools, technologies, and applications. *Microbiol Mol Biol Rev* **68**(3):538–559

[104] Wagner M (2009) Single-cell ecophysiology of microbes as revealed by Raman microspectroscopy or secondary ion mass spectrometry imaging. *Annu Rev Microbiol* **63**:411–429

[105] Davis KM and Isberg RR (2016) Defining heterogeneity within bacterial populations via single cell approaches. *BioEssays* **38**(8):782–790

[106] Kreft JU, Booth G, and Wimpenny JWT (1998) BacSim, a simulator for individual-based modelling of bacterial colony growth. *Microbiology* **144**(12):3275–3287

[107] Lardon LA, Merkey BV, Martins S, Dötsch A, Picioreanu C, Kreft JU *et al.* (2011) iDynoMiCS: Next-generation individual-based modelling of biofilms. *Environ Microbiol* **13**(9):2416–2434

[108] Jayathilake PG, Gupta P, Li B, Madsen C, Oyebamiji O, González-Cabaleiro R *et al.* (2017) A mechanistic individual-based model of microbial communities. *PLoS ONE* **12**(8):e0181965

[109] Varma A and Palsson BØ (1994) Stoichiometric flux balance models quantitatively predict growth and metabolic by-product secretion in wild-type *escherichia coli* w3110. *Appl Environ Microbiol* **60**(10):3724–3731

[110] Kauffman KJ, Prakash P, and Edwards JS (2003) Advances in flux balance analysis. *Curr Opin Biotechnol* **14**(5):491–496

[111] Raman K and Chandra N (2009) Flux balance analysis of biological systems: applications and challenges. *Brief Bioinform* **10**(4):435–449

[112] McCloskey D, Palsson BØ, and Feist AM (2013) Basic and applied uses of genome-scale metabolic network reconstructions of *Escherichia coli*. *Mol Syst Biol* **9**

[113] Palsson BØ (2015) *Systems biology: constraint-based reconstruction and analysis.* Cambridge University Press. ISBN 1107038855

[114] Schuetz R, Kuepfer L, and Sauer U (2007) Systematic evaluation of objective functions for predicting intracellular fluxes in *Escherichia coli. Mol Syst Biol* **3**(119)

[115] Feist AM and Palsson BØ (2010) The biomass objective function. *Curr Opin Microbiol* **13**(3):344–349

[116] Thiele I and Palsson BØ (2010) A protocol for generating a high-quality genome-scale metabolic reconstruction. *Nat Protoc* **5**(1):93–121

[117] Ravikrishnan A and Raman K (2015) Critical assessment of genome-scale metabolic networks: the need for a unified standard. *Brief Bioinform* **16**(6):1057–1068

[118] Henry CS, DeJongh M, Best AA, Frybarger PM, Linsay B, and Stevens RL (2010) High-throughput generation, optimization and analysis of genome-scale metabolic models. *Nat Biotechnol* **28**:977–982

[119] Caspi R, Altman T, Dreher K, Fulcher CA, Subhraveti P, Keseler IM *et al.* (2012) The MetaCyc database of metabolic pathways and enzymes and the BioCyc collection of pathway/genome databases. *Nucleic Acids Res* **40**(Database issue):D742–D753

[120] Kanehisa M, Sato Y, Kawashima M, Furumichi M, and Tanabe M (2016) KEGG as a reference resource for gene and protein annotation. *Nucleic Acids Res* **44**(Database issue):D457–D462

[121] King ZA, Lu J, Dräger A, Miller P, Federowicz S, Lerman JA *et al.* (2016) BiGG models: A platform for integrating, standardizing and sharing genome-scale models. *Nucleic Acids Res* **44**(Database issue):D515–D522

[122] Büchel F, Rodriguez N, Swainston N, Wrzodek C, Czauderna T, Keller R *et al.* (2013) Path2Models: large-scale generation of computational models from biochemical pathway maps. *BMC Syst Biol* **7**:116

[123] Zomorrodi AR and Maranas CD (2012) OptCom: A multi-level optimization framework for the metabolic modeling and analysis of microbial communities. *PLoS Comput Biol* 8(2):e1002363

[124] Zomorrodi AR, Islam MM, and Maranas CD (2014) d-OptCom: Dynamic multi-level and multi-objective metabolic modeling of microbial communities. *ACS Synth Biol* 3(4):247–257

[125] Khandelwal RA, Olivier BG, Röling WFM, Teusink B, and Bruggeman FJ (2013) Community flux balance analysis for microbial consortia at balanced growth. *PLoS ONE* 8(5):e64567

[126] Zelezniak A, Andrejev S, Ponomarova O, Mende DR, Bork P, and Patil KR (2015) Metabolic dependencies drive species co-occurrence in diverse microbial communities. *Proc Natl Acad Sci U S A* 112(20):6449–6454

[127] Mahadevan R, Edwards JS, and Doyle FJ (2002) Dynamic flux balance analysis of diauxic growth in *Escherichia coli*. *Biophys J* 83(3):1331–1340

[128] Hanly TJ and Henson MA (2011) Dynamic flux balance modeling of microbial co-cultures for efficient batch fermentation of glucose and xylose mixtures. *Biotechnol Bioeng* 108(2):376–385

[129] Zhuang K, Izallalen M, Mouser P, Richter H, Risso C, Mahadevan R *et al.* (2011) Genome-scale dynamic modeling of the competition between *Rhodoferax* and *Geobacter* in anoxic subsurface environments. *ISME J* 5(2):305–316

[130] Harcombe WR, Riehl WJ, Dukovski I, Granger BR, Betts A, Lang AH *et al.* (2014) Metabolic resource allocation in individual microbes determines ecosystem interactions and spatial dynamics. *Cell Rep* 7(4):1104–1115

[131] Hanly TJ and Henson MA (2013) Dynamic metabolic modeling of a microaerobic yeast co-culture: predicting and optimizing ethanol production from glucose/xylose mixtures. *Biotechnol Biofuels* 6(1):44

[132] Zou W, Liu L, and Chen J (2013) Structure, mechanism and regulation of an artificial microbial ecosystem for vitamin C production. *Crit Rev Microbiol* **39**(3):247–255

[133] Du J, Zhou J, Xue J, Song H, and Yuan Y (2012) Metabolomic profiling elucidates community dynamics of the *Ketogulonicigenium vulgare–Bacillus megaterium* consortium. *Metabolomics* **8**:960–973

[134] Ding MZ, Zou Y, Song H, and Yuan YJ (2014) Metabolomic analysis of cooperative adaptation between co-cultured *Bacillus cereus* and *Ketogulonicigenium vulgare*. *PLoS ONE* **9**(4):e94889

[135] Zou W, Liu L, Zhang J, Yang H, Zhou M, Hua Q *et al.* (2012) Reconstruction and analysis of a genome-scale metabolic model of the vitamin C producing industrial strain *Ketogulonicigenium vulgare* WSH-001. *J Biotechnol* **161**:42–48

[136] Zou W, Zhou M, Liu L, and Chen J (2013) Reconstruction and analysis of the industrial strain *Bacillus megaterium* WSH002 genome-scale *in silico* metabolic model. *J Biotechnol* **164**:503–509

[137] Rubbens P, Props R, Boon N, and Waegeman W (2017) Flow cytometric single-cell identification of populations in synthetic bacterial communities. *PLoS ONE* **12**(1):e0169754

[138] Fakruddin M and Mannan KSB (2013) Methods for analyzing diversity of microbial communities in natural environments. *Ceylon J Sci Biol Sci* **42**(1):19–33

[139] De Roy K, Clement L, Thas O, Wang Y, and Boon N (2012) Flow cytometry for fast microbial community fingerprinting. *Water Res* **46**(3):907–919

[140] Props R, Monsieurs P, Mysara M, Clement L, and Boon N (2016) Measuring the biodiversity of microbial communities by flow cytometry. *Methods Ecol Evol* **7**(11):1376–1385

[141] Prest EI, Weissbrodt DG, Hammes F, Van Loosdrecht MC, and Vrouwenvelder JS (2016) Long-term bacterial dynamics in a full-scale drinking water distribution system. *PLoS ONE* **11**(10):e0164445

[142] Shaikh AS, Tang YJ, Mukhopadhyay A, and Keasling JD (2008) Isotopomer distributions in amino acids from a highly expressed protein as a proxy for those from total protein. *Anal Chem* **80**(3):886–890

[143] Rühl M, Hardt WD, and Sauer U (2011) Subpopulation-specific metabolic pathway usage in mixed cultures as revealed by reporter protein-based ^{13}C analysis. *Appl Environ Microbiol* **77**(5):1816–1821

[144] Ghosh A, Nilmeier J, Weaver D, Adams PD, Keasling JD, Mukhopadhyay A *et al.* (2014) A peptide-based method for ^{13}C metabolic flux analysis in microbial communities. *PLoS Comput Biol* **10**(9):e1003827

[145] Bordbar A, Lewis NE, Schellenberger J, Palsson BØ, and Jamshidi N (2010) Insight into human alveolar macrophage and *M. tuberculosis* interactions via metabolic reconstructions. *Mol Syst Biol* **6**:422

[146] Huthmacher C, Hoppe A, Bulik S, and Holzhutter HG (2010) Antimalarial drug targets in *Plasmodium falciparum* predicted by stage-specific metabolic network analysis. *BMC Syst Biol* **4**:120

[147] Sinha S, Grewal RK, and Roy S (2018) *Modeling Bacteria–Phage Interactions and Its Implications for Phage Therapy*, volume 103, pp. 103–141. Elsevier. ISBN 9780128151839

[148] Samaddar S, Grewal RK, Sinha S, Ghosh S, Roy S, and Gupta SKD (2015) Dynamics of Mycobacteriophage-Mycobacterial host interaction: Evidence for secondary mechanisms for host lethality. *Appl Environ Microbiol* **82**(1):124–133

[149] Heinken A and Thiele I (2015) Systematic prediction of health-relevant human-microbial co-metabolism through a computational framework. *Gut Microbes* **6**(2):120–130

[150] Karr JR, Sanghvi JC, Macklin DN, Gutschow MV, Jacobs JM, Bolival B *et al.* (2012) A whole-cell computational model predicts phenotype from genotype. *Cell* **150**(2):389–401

[151] Szigeti B, Roth YD, Sekar JAP, Goldberg AP, Pochiraju SC, and Karr JR (2018) A blueprint for human whole-cell modeling. *Curr Opin Syst Biol* **7**:8–15

Index

Printed and bound by CPI Group (UK) Ltd, Croydon, CR0 4YY

22/10/2024

01777627-0001